STEAM Laboratory Activities for the Classroom

FOCUS ON CHEMISTRY

Stephen J. Wolf, PhD

Print Edition:
ISBN: 978-1-950110-52-0

To all my students through the years. Thank you for inspiring me.

Table of Contents

INTRODUCTION

Purpose
This manual is intended for teachers to use in their middle or high school classes to further student involvement in science, technology, engineering, art, and mathematics (STEAM). The lessons in this book focus on chemistry experiments.

Homeschooling
Most of these labs can be used for homeschooling. You may want to invest in a food scale to use in place of a digital balance. Regular glassware can be substituted for beakers, and a measuring cup can often serve the purpose of a graduated cylinder. Safety goggles can be found at a local hardware store.

Research
Science also involves research into the information that is already known so scientists can make informed decisions on their experiments. To this end, there are some research tasks included for students to complete linked to certain experiments. Some of these involve answering a series of questions, while others involve writing a longer essay.

Materials
Safety is of utmost priority. Therefore, the majority of materials are safe for younger student use. Some activities call for the use of hot plates. When a materials list is given for an activity, it will be for one group of students, unless noted otherwise. Make the necessary calculations to serve your class.

Lab Discussions
For all labs, I recommend walking students through the procedure before starting. The labs come with analysis questions for the students to answer. These questions should be reviewed in a class discussion once the lab activity has been completed, or possibly the next day.

Reproductions of text
Because this text is expected to be used in classrooms, teachers may make copies of the lab sheets as needed. However, if multiple teachers are utilizing the labs, please purchase additional copies of this manual by navigating through red-jade.com/steam.

About the Author

Stephen J. Wolf has a PhD in Science Education and is a science teacher with 22 years of experience working with middle and high school students. His background is in chemistry and physics.

When he was a child, Dr. Wolf was enamored with Mr. Wizard's World, where science experiments were used to explore the world and explain mysteries of why things work the way they do. Science was a way to explain "magical" events, such as why fireworks light different colors when they explode. It didn't take long for him to get hooked.

Dr. Wolf also engaged in computer programming at various points in his life, from coding programs on the Commodore 64 in BASIC to more recent dives into JavaScript. Dr. Wolf has written three *Coding for Kids* books geared to teaching students the fundamentals of creating interactive webpages and games. For more information, be sure to check out red-jade.com/coding.

Dr. Wolf is also a creative writer, focusing on fantasy. He authored a short story, *A Shocking Journey*, geared toward middle school students to teach the concepts of electricity and magnetism through the use of analogies. Each chapter focuses on a different aspect of electricity and magnetism, such as the structure of the atom, electron energy, and series and parallel circuits. For more information, visit red-jade.com/shocking.

Because Dr. Wolf has a penchant for fantasy, he has also written several fantasy books, including his *Red Jade* series, the *Starsword* series, *Kershin the Fire Mage*, and a collection of *Choose-the-Fate* stories.

Dr. Wolf lives in New York with his husband, Kevin, and two cats, Merlin and Monty. You can visit his blog at StephenJWolf.com or learn more about his writings at Red-Jade.com.

CHAPTER 1

The Scientific Method

To explore laboratory exercises in a focused, concise manner, it is important to utilize the process known as the Scientific Method. This is a six-step process that we use all the time without realizing it. Our brains are already wired to follow this process and this chapter is meant to slow us down and examine the steps we need to properly set up and control our experiments.

The following pages can be handed directly to students, but it may be worthwhile converting the text to a slideshow to focus on one concept at a time.

Scientific Method Discussion

The scientific method is a six-step process. It allows us to examine a problem in an organized way. There are a lot of details here, but as we use this process, these details will become automatic. Let's talk about it.

STEP 1 – IDENTIFY THE PROBLEM

- This defines the topic you need to solve
- The problem needs to be in the form of a question
- The question needs to be open-ended, not a yes/no question
- Ex: How does adding salt affect the density of water?

STEP 2 – DO SOME RESEARCH

- This can be from brainstorming what you already know about the problem
- You can also find information from books, the internet, etc.
- Ex: salt dissolves in water

STEP 3 – COME UP WITH YOUR HYPOTHESIS

- This is an educated guess or prediction about the outcome
- It's "educated" because it is based on the research
- This must be in the form of a statement
- "As this (independent variable) happens, that (dependent variable) will happen."
- Use the variable names in your hypothesis (variables are discussed below)
- Ex: Adding salt will increase the density of water

STEP 4 – CONDUCT YOUR EXPERIMENT

- Collect data in an organized way
- There are many steps to properly controlling an experiment

PROCEDURE

- This is a set of detailed steps needed to perform an experiment
- This should be clear enough that anyone can follow your procedure without needing help
- Ex:
    1. Obtain the mass of a container.
    2. Fill the container with 100 mL of water.
    3. Obtain the mass of the container with the water.
    4. Subtract the two values to find the mass of the water.
    5. Calculate the density of the water using the following formula.

$$density = \frac{mass}{volume}$$

        - Take the mass of the water and divide it by the volume of water.
        - Keep two decimal places.
    6. Obtain the mass of another container.
    7. Add salt until the mass has increased by 10 grams.
    8. Add the salt to the water and stir.
    9. Obtain the mass of the saltwater and container
    10. Subtract out the mass of the original container to obtain the mass of the saltwater.
    11. Calculate density as you did in Step 5.
    12. Repeat Steps 6 through 11 for two more trials.
    13. Clean up your lab station.

INDEPENDENT VARIABLE (IV)

- "I" change this; changed by the experimenter
- This is the variable being tested
- It's what you want to learn the effect of
- A properly controlled experiment only has one IV
    - More detailed experiments can have more independent variables
        - This requires more trials to be performed.
- Ex: amount of salt being added

DEPENDENT VARIABLE (DV)

- "D"ata being collected
- This changes as a result of the independent variable
- You may have several DVs in an experiment
- Ex: density of the saltwater solution

CONSTANTS

- Sometimes called "controlled variables"
- These are possible variables that you control to remain the same
- They remain the same for all trials in an experiment
- Do not list tools such as a stopwatch or meter stick
- Always try to list three constants
- *Think*: If you did not control this, it would change the nature of the experiment
- Ex: temperature of the water, type of salt, not letting the salt dissolve

CONTROL GROUP

- "No" or "Normal" group
- This is a group (trial) used for comparison
- This is not experimented upon; it does not receive treatment
    - In some experiments, this is impossible, so then use your first trial as your control group
- This is one trial of the independent variable
- Ex: water without any salt added

EXPERIMENTAL GROUP

- These are groups that are part of the actual experiment
- These involve the changing of the independent variable
- These are the other trials of the independent variable
- Ex: the trials where salt was added

TYPES OF DATA

- There are two main classes of data
- Qualitative data
    - This is non-numerical data that describes a situation in words
    - Some qualitive data can be converted into quantitative data by use of a Likert scale
        - 1 – very little dissolved, 2 – some dissolved, 3 – a lot dissolved, 4 – all dissolved
        - This allows you to graph qualitative data, which normally can't be graphed
    - Ex: salt is a white, crystalline powder that dissolves when it's stirred in water
- Qualitative data
    - This is numerical, measurable data
    - Generally, this is easy to graph and can be used to find patterns or trends
    - Ex: 10 grams of salt dissolved in water produced a density of 1.05 g/mL

DATA TABLE

- Record all data in a chart
- Make the first column the trial number
- Make the second column the independent variable
  - If you have more than one independent variable, the data table will be more complex
- Make the third column the dependent variable
  - If you have more than one dependent variable, add more columns

Example with one IV:

Trial #	Independent Variable Amount of Salt	Dependent Variable Density
Trial 1 (Control Group)	0 grams	1.00 g/mL
Trial 2 (Experimental Group)	10 grams	1.05 g/mL
Trial 3 (Experimental Group)	20 grams	1.12 g/mL
Trial 4 (Experimental Group)	30 grams	1.21 g/mL

Example with two IVs:

Trial #	IV: Temperature: Cold Water		IV: Temperature: Warm Water	
	IV: Amount of Salt	DV: Density	IV: Amount of Salt	DV: Density
1 (Control)	0 grams	1.00 g/mL	0 grams	0.90 g/mL
2 (Experimental)	10 grams	1.05 g/mL	10 grams	0.96 g/mL
3 (Experimental)	20 grams	1.12 g/mL	20 grams	1.01 g/mL
4 (Experimental)	30 grams	1.21 g/mL	30 grams	1.15 g/mL

STEP 5 – DO THE ANALYSIS

- Look at your data
- Make a graph
- Try to find trends or patterns
- Example with one IV:

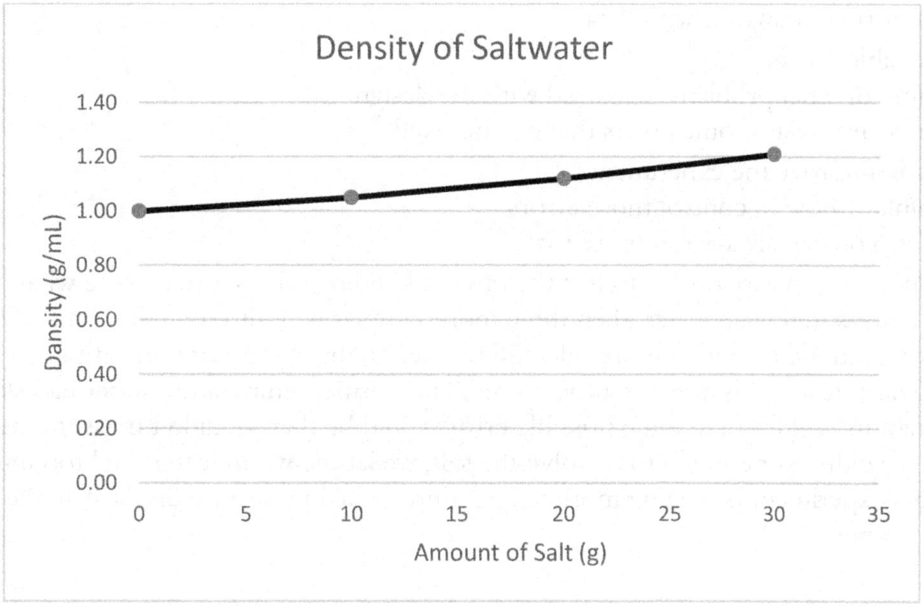

- *Trend*: As more salt was added, the density increased.

- Example with two IVs:

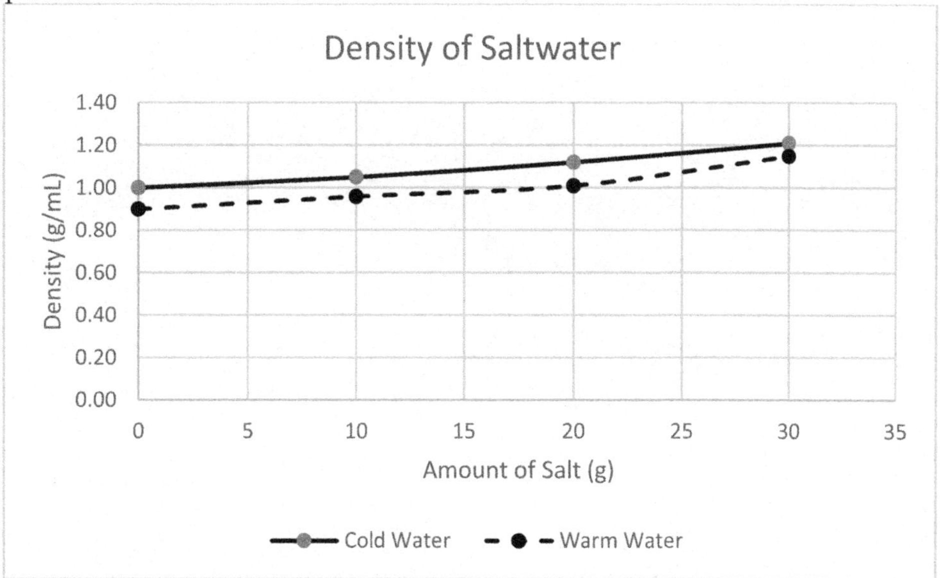

- *Trends*: As more salt was added, the density increased. Also, warm water had a lower density.

STEP 6 – WRITE A CONCLUSION

- Briefly review what you did
- Refer back to your hypothesis
  - Was it right or wrong?
    - If it was wrong, that's okay! Don't change it. You still learned something.
  - Support your answer with data
- Discuss possible errors
  - Think of what problems occurred with the design
  - There are always some errors that can be fixed
- How can you improve the experiment?
  - Think of how to control those errors
  - *Hint*: You can always run more trials
- Ex: We conducted an experiment to test the effect of adding salt to water to see what happens to density. We measured plain water, then three more samples with different amounts of salt. We thought the density would increase with more salt, and we were right. As the amount of salt increased, so did the density. We also tested this at two temperatures. The warmer temperature water had slightly lower densities than the cold water. Some possible errors could be that we didn't measure the salt correctly each time; we didn't stir enough to dissolve the salt; we let the warm water cool too much. We can improve this experiment by testing more temperatures, making sure we dissolve all the salt, and by trying other types of salt.

Final Thoughts

The scientific method is a detailed process, but we use it all the time. Follow these steps and you're sure to have a well-controlled experiment.

Lab Safety

Safety is of utmost importance in the science classroom. Without accounting for safe practices, people can get hurt, and this can be a liability for the teacher, especially in terms of neglect. Be sure to conduct a lab safety lesson with the students and have them and their parents sign and return a lab safety contract. There are many safety contracts online, such as on the Flinn Scientific site, but they are generally focused on high school laboratory settings.

On the next page is an example safety contract. Be sure your chairperson approves its use if you choose to utilize it. There are two copies on the sheet. I collect one and let the student keep the other.

This should not simply be handed out to students to sign. Time should be spent discussing the importance of following safety steps in the classroom. I recommend converting the lab safety contract into a slideshow to review with the students before handing out the contract.

Middle School Lab Safety Contract

Ways to be safe

- read and follow directions carefully
- keep glassware away from edges
- don't leave round objects lying on counters
  - test tubes, thermometers, etc.
- no fooling around, pushing, or shoving
- be aware of others so you don't bump into them
- keep the aisles clear
- no food or drink is allowed in the lab
- goggles must be worn when using hot plates
- **NEVER** touch materials on a hot plate

NEVER TOUCH ANOTHER PERSON'S EXPERIMENT, no matter how curious you are. You could ruin their data, break something, or get exposed to dangerous chemicals

Know the locations of safety equipment.

- Fire extinguishers, fire blanket, eyewash
- Do not use without permission

If something breaks, tell your teacher.

- Your teacher will clean up broken glassware
- Don't do it yourself. You may get cut.

If something spills, tell your teacher.

- Most (if not all) the chemicals we use will be safe to touch
- Listen for special instructions
- Clean up safe spills with a sponge

Clean up

- Your station MUST be clean before you leave. Other classes use this room.

Student Name____________________________

Student Signature_________________________

Parent Signature__________________________

Period___ Date_______ Teacher____________

Middle School Lab Safety Contract

Ways to be safe

- read and follow directions carefully
- keep glassware away from edges
- don't leave round objects lying on counters
  - test tubes, thermometers, etc.
- no fooling around, pushing, or shoving
- be aware of others so you don't bump into them
- keep the aisles clear
- no food or drink is allowed in the lab
- goggles must be worn when using hot plates
- **NEVER** touch materials on a hot plate

NEVER TOUCH ANOTHER PERSON'S EXPERIMENT, no matter how curious you are. You could ruin their data, break something, or get exposed to dangerous chemicals

Know the locations of safety equipment.

- Fire extinguishers, fire blanket, eyewash
- Do not use without permission

If something breaks, tell your teacher.

- Your teacher will clean up broken glassware
- Don't do it yourself. You may get cut.

If something spills, tell your teacher.

- Most (if not all) the chemicals we use will be safe to touch
- Listen for special instructions
- Clean up safe spills with a sponge

Clean up

- Your station MUST be clean before you leave. Other classes use this room.

Student Name____________________________

Student Signature_________________________

Parent Signature__________________________

Period___ Date_______ Teacher____________

Possible Mission – Teacher Notes

In this exercise, you're going to challenge the students with a task that seems impossible. However, it *can* be completed within the restrictions. It is recommended that you practice the solution until you know it firmly so you can offer guidance to the students as they try to work through it. The task is to cut up a 3"x5" index card so one student can fit through it without fastening it in any way.

Possible Mission Lab

Time: One class period

Note: there is no lab sheet for this activity

Materials per group

- One pair of scissors
- About 10 3"x5" index cards

To the class

Today, I'm going to offer you a tricky challenge. It's a task that will seem impossible, but I assure you that you can do it. There is one major rule that you must follow, and I'll get to that in a moment. I need you to put your thinking caps on and to be creative. You'll be working in pairs [or groups] to solve the mission.

Your challenge is to fit one member of your group through… [pull out one index card] … an index card. [Wait a moment for the students to comment and settle. It's possible some may have seen this before. If so, ask them not to reveal the answer to their group, but to offer guidance instead.]

You are only allowed one pair of scissors. You may NOT fasten the card in any way. This includes using tape or staples. You may not hold two pieces together. You may not tie two ends together. You need to create one continuous loop. There are no gimmicks. You cannot write your name and pass it through the card. You can't just stick your finger through it either. Your whole body has to pass through.

It sounds tough and impossible, but I promise you that it can be done. I want you to brainstorm in your group how to solve this problem. When you have an idea, I will give you your first index card. If you make a mistake and need a replacement, let me know and I'll give you another one.

Solution

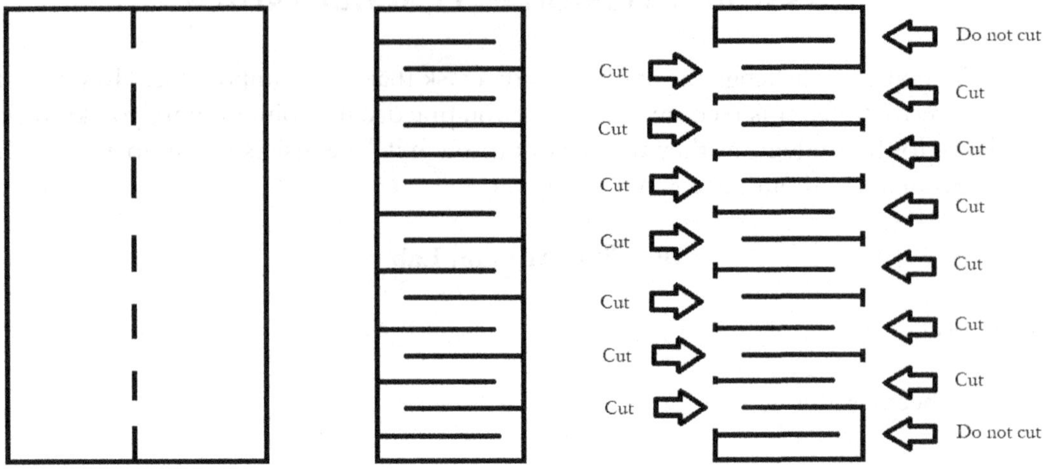

Fold in half lengthwise

Make alternating cuts from each side

Cut off the folded edges, but not the topmost or the bottommost

Comments

Students are likely to be frustrated as they work through this, so it's important to walk around and offer encouragement. Students will often come up with a spiral, which is a great idea and along the lines of being correct. They could actually convert the spiral into a loop by slicing the inside of the spiral from end to end without cutting the very ends.

Most students will try to cheat the system by cutting a long spiral and holding or tying the ends together. That's a no-no. It breaks the rule of no fastening. Some students will cut small strips and set it in a circle. Also, this doesn't count. But in both cases, the students should be supported for their ingenuity and reminded of the rule of no fastening.

When running this with my students, about halfway into the activity, I take an index card and start cutting it into the solution. Students will see this and try to replicate it, which is good – scientists often work together and take inspiration from and observations of the world around them. At the end, I open the loop and work it over my head and down to my feet. It's tricky doing this without it breaking. The closer the cuts are made to each other, the larger the loop will be, but it will also be more fragile.

My last talking point to the students is that I won't ever give them something that can't be done. Everything we work on is possible with some effort and sticking to the scientific method and our procedures.

Prediction/Observation – Teacher Notes

Students will mix together food coloring, vegetable oil, and water to create a dazzling effect. This lab touches a little on molecular polarity and density, if you're inclined to go into those topics at this time. We'll be covering density in a bit more detail in an upcoming chapter, but I'm giving background here in case you want to discuss it now. This lab can be messy, especially during the filtration step. I recommend having all the work done on trays, if available.

Time: One class period

Materials per group

- Two 250 mL beakers
- 100 mL water
- 75 mL vegetable oil
- Food coloring (4 colors)
- Stirring stick (spoons can make a mess)
- Coffee filter
- Work tray (if available)
- Small amount of dish soap (for cleanup)

Additional Material

- Plastic bottle(s) for the used vegetable oil for disposal

Background

Density is a property of matter that tells us how close molecules are in a substance. It's related to the mass (the amount of material) and the volume (the amount of space it occupies). The size of the sample doesn't matter, only the closeness of the molecules. Density is useful because it can predict whether something will float on something else. If two liquids are combined, the one with the lower density will float on top.

Molecular polarity helps determine whether two substances can mix together. Molecules can be *polar* or *nonpolar*. A water molecule has a bent shape that gives it slightly polar ends. Oil molecules are large and balanced and are nonpolar. When you put these two together, you can't mix them. If you stir them vigorously, they do mix for a while, but they separate again. Soap is a special material that has a polar end and a nonpolar end. It's able to bring these two materials together, and that's why we need to use soap to clean up oil if we spill it.

For the teacher

Don't announce this to the students, but food coloring is mostly water, therefore it is polar and will not mix with the oil. It will become encapsulated in an oil bubble and sink to the bottom of the beaker. Stirring will break the food coloring blobs apart into smaller bubbles. This usually fascinates the kids. When pouring the oil onto the water, it is important to stir again so the color blobs are floating in the oil first. The students should watch the combination from eye level. After a minute or so, the food coloring will breach the water level and spread out like colored lightning. At this time, you should pour off most of the oil into the waste bottle, leaving the colored water behind. Students will then filter the water and should see the colors separate out again on the coffee filter.

To the class

Today we're going to perform a prediction/observation lab. At each step, you must make a prediction ***before*** you carry out the step. It's easy to forget to make the prediction first, but it's important. [Note: most kids will forget.] For example, in Part A, you'll predict what happens when you add food coloring to oil. You may be tempted to just add the colors, but you need to write your guess first. Also, you may be tempted to stir it. Don't! That's for Part B.

This lab can be a bit messy if you're not careful. Keep all your lab materials on your trays [if you have them]. At the end, when it's time to clean up, you need to use a drop or two of dish soap to clean the oil out of the beakers and to clean your workstation. We're dealing with glassware, so take care not to break anything. If you do, call me over immediately. Do not clean up broken glass on your own.

Comments

Students are likely to squirt in a bunch of food coloring, so you may want to emphasize that it's only 2-3 drops ahead of time. Also, when it's time to filter the food coloring and water, be sure to watch how they set up the filter. Some will hold it tightly over the other beaker, which will make a mess. The filter needs to bow into the beaker before and during pouring. It's helpful for one student to hold the filter while another pours.

Most likely, the students will struggle to clean the oil properly.

After this lab, there is an opportunity to use a class period to discuss the results with the students. I have the discussion section set up as a slideshow that we talk through together. Attached after the lab is the information from the slideshow.

Names Period Date Teacher

Prediction/Observation Lab

Purpose: In this lab, you will explore the effects of food coloring on oil.

Materials: 100 ml water, 75 ml vegetable oil, food coloring, coffee filter, stirrer, two 250 ml beakers

Procedure: See the steps below. Read and follow them carefully and in order.

Part A: Food Coloring

1. **Predict**. What will happen when you put a few drops of food coloring into the oil?

2. **Explore**. Once you have written your prediction, choose a color and add **2-3 drops** of food coloring to the oil.

3. **Observe**. What happened to the food coloring? Was this what you expected?

4. **Explore**. Add the other three colors to the mixture and note what happens. Do the colors mix?

Part B: Stirring

5. **Predict**. What will happen when you stir the mixture?

6. **Explore**. Take a stirring stick and vigorously stir the mixture. Hold the beaker. Take care not to spill.

7. **Observe**. What happened to the mixture? Was it what you predicted?

Part C: Combination

8. **Predict.** Next, you will pour the mixture into a beaker of water. What do you think will happen?

9. **Explore.** Take the beaker with the mixture of oil and stir it. Then pour it into the beaker of water. (Don't do it the other way.) Let the mixture sit for a few minutes before moving on. Keep your eyes on it.

10. **Observe.** Describe what happened to the mixture. Was it what you predicted?

Part D: Coffee Filter

11. **Predict.** What will happen if you pour the mixture through a coffee filter?

12. **Explore.** First, ask your teacher to pour off the excess oil. Place the coffee filter over the empty beaker. Make sure it forms a "cup" inside the beaker so you can pour into it without spilling. Then, SLOWLY pour the mixture into the filter.

13. **Observe.** What happened to the mixture? What does the coffee filter look like? Did this match your prediction?

Part E: Clean Up

14. **Clean up**
    - Discard the coffee filter into the garbage. Take care not to drip on the floor.
    - Bring your beakers to the sink and add ONE DROP of soap to each, then scrub and rinse each out until there is no oil left.
    - Hang the beakers on the drying rack. You don't need to dry them with paper towels.
    - Wash and dry your hands.

15. **Reflection.** Think about the lab experience and reflect on what you did.
    - What were the best and worst parts of this lab?
    - What could be changed about this lab to make it better?

Prediction/Observation Lab – Discussion

Problem

What would be a good question to focus this lab?

- How does food coloring affect oil?
- What happens when food coloring is added to oil?
- What happens when the mixture is added to water?

Research

What did you already know before you started? Or, what did you research before beginning?

- Food coloring changes the color of things
- Oil and water don't mix
- Oil floats on water
- *Food coloring is water-based

Why does oil float on water and not water on oil?

- vegetable oil has a density of about 0.86 g/mL
- water has a density of 1.00 g/mL
- because the oil has a lower density, it floats on top

Why don't oil and water mix?

- they have different polarities
- water is polar (has a relative charge)
- oil is nonpolar (does not have a charge)
- different polarities do not mix

Hypothesis

What was the guess you made for Part A (adding food coloring)?

- The colors would mix
- The colors would not mix
- Other?

What was the guess you made for Part B (stirring)?

- The colors would mix with the oil
- The colors would split into smaller pieces
- Other?

What was the guess you made for Part C (combining with water)?

- The mixture would turn a different color
- The oil and water would mix
- Other?

What was the guess you made for Part D (filtering)?

- The colors would all go through the paper
- The colors would separate out
- Other?

Experiment: Procedure

What steps did you follow?

- For all parts, predict first, then carry out instructions
- Part A:
    - Make a prediction
    - Choose one color
    - Add two drops of the color to the oil mixture
    - Observe
    - Repeat with other colors
- Part B:
    - Make a prediction
    - Stir vigorously
    - Observe
- Part C:
    - Make a prediction
    - Pour the oil mixture into the water beaker
    - Observe
- Part D:
    - Make a prediction
    - Pour off the excess oil
    - Set up the filter over the other beaker
    - Pour the mixture slowly into the filter paper
    - Observe the filter paper

Experiment: Independent Variable

What part of the experiment did you have control over in Part A?

- Color added

Experiment: Dependent Variable

What data did you gather in Part A?

- Whether or not the colors mixed with the oil

Experiment: Constants

What are possible variables that were kept the same and were controlled not to change in Part A?

- Added the same amount of each color
- Liquids were all room temperature
- No stirring for any trial

Experiment: Control Group

What was the comparison group for Part A of this experiment?

- Because all trials received the independent variable, choose one trial as the comparison
- Whatever your first color was, such as red

Experiment: Experimental Groups

What were the experimental groups for Part A of this experiment?

- The other colors you added, such as yellow, blue, and green

Experiment: Data Type

What type of data was collected, qualitative or quantitative?

- Qualitative
- We didn't measure any numerical data

Experiment: Data Table

What would a good data table for Part A look like? (We didn't make one this time.)

Group	Independent Variable	Dependent Variable
	Color Added	**Did it Mix with the Oil?**
Control	Red	No
Experimental	Yellow	No
Experimental	Blue	No
Experimental	Green	No

This data table isn't very informative.

Analysis

What did the data say? What did you learn? Was there a trend or pattern?

- Trend: None of the liquids mixed with oil

Conclusion: Summary

Write a brief description of what you did.

- We added four different colors of food coloring to a beaker of oil. We then stirred the mixture. We poured the mixture into a beaker of water. Finally, we separated the mixture with a coffee filter.

Conclusion: Hypothesis

Was your hypothesis (prediction) correct?

- In Part A, our hypothesis was wrong. We thought it was going to mix, but it didn't. Then we were right about the other colors.
- In Part B, our hypothesis was partly right. The colors didn't mix when we stirred, but they broke into smaller parts.
- In Part C, our hypothesis was wrong. We didn't expect the colors to seep through into the water like they did.
- In Part D, our hypothesis was right, because the colors separated out onto the coffee filter.

Conclusion: Errors

What are possible errors that happened during the experiment? Try to list three.

- We accidentally put more than two drops
- We spilled the oil
- We broke a beaker
- We didn't measure the liquids carefully
- We didn't stir enough
- We didn't wait long enough to make observations

Experiment: Improvements

How can the results for this experiment be improved? Generally, choose one error and discuss how to fix it. But you can come up with other ideas too.

- Measure carefully
- Don't break glassware
- Try the experiment with different liquids
- Try the experiment with different temperatures

The End

That was a lot of detail for a simple experiment, but it's important to be able to break down your work into individual steps. These steps help make sure you have a good, properly controlled experiment.

Properties of Matter – Teacher Notes

In this lab, students will examine through comparisons various materials to determine what properties objects have. There is some setup involved, but once it's set, you can easily save and store it for future lessons. I generally set up two sets of stations so groups can be smaller. You can certainly substitute with other items for each station, just make sure at least one property (i.e. size) is consistent within the station's materials. Label each item A through E at each station in any order, but be consistent if you're setting up more than one station.

__Time__: One class period

__**Materials per station**__
Station 1 (same length): plastic knife, metal spatula, rubber band (cut), drinking straw, string
Station 2 (cubes): copper cube, iron cube, aluminum cube, large die, paper cube
Station 3 (cylinders): copper cylinder, iron cylinder, aluminum cylinder, brass cylinder, paper cylinder
Station 4 (squares): aluminum foil, wax paper, cardboard, clear plastic wrap, cloth

__**Additional optional materials**__
Rulers, balances

__**To the class**__

> Today you will be using your powers of observation to determine various properties of matter. At each station, compare and contrast the materials you see. Write down common features and different features. We're not concerned about what they're used for, just what they look and feel like. There are five objects at each station. Make sure none fall on the floor. You will have about four minutes per station.

__**Pre-lab example**__
To demonstrate what to do, tape up five samples of circles and rectangles (and a square) on the board of different sizes and colors. Have the students give you descriptions of each circle. [no angles, no beginning or end, etc.] Ask, "Does the size matter? [no] This will be the same for our lab." Now repeat for the rectangles.

At the end of the investigation, bring the students together to discuss their findings. Focus on one station at a time. These are some suggestions for notes.

__**Physical properties by station**__

Station 1:
Flexibility: ability of a material to bend (plastic straw, string)
Elasticity: ability of a material to stretch (rubber band)

Station 2:
Color: depends on reflected light (not absorbed light)

Station 3:
Metallic character (Metallic property): shows how metallic a material is (paper versus copper)

Station 4:
Luster (Metallic property): how shiny something is (aluminum foil)
Opaque: light cannot pass through (cardboard)
Translucent: some light can pass through (wax paper)
Transparent: all light can pass through (plastic wrap)
Texture: relative smoothness (cloth versus aluminum foil)

Cross-station observations
Shape: physical arrangement of molecules (station 2 compared to station 3)
Phase: solid, liquid, or gas (all materials were solids)
Malleability (Metallic property): ability to hammer into sheets (aluminum block vs foil)

Additional physical properties to discuss
Density: closeness of molecules in a substance (in this lab, related to mass since volumes were the same, but this is a common misconception, that mass equals density, so I don't usually point this out at this time.)
Hardness: material's resistance to scratching (Mohs scale: diamond is the hardest natural mineral, talc is the softest)
Conductivity (Metallic property): ability to draw electricity through it (copper has high conductivity, cloth and rubber have none)
Ductility (Metallic property): ability to be pulled into wires (copper is very ductile)
Magnetism (Metallic property): ability to be drawn by a magnet (iron, cobalt, and nickel are naturally strong magnets)
Melting point: temperature at which something freezes or melts
Boiling point: temperature at which something condenses or boils
Solubility: ability of a material to dissolve in a given solvent

Those were *physical properties* (how things look and feel).

Physical Change: altering the physical state of an object, does not create new compounds
Heating or cooling to change the phase, breaking a window, scratching a surface, cutting a material

Here are a couple of *chemical properties* (how things react or interact).

Chemical Properties
Flammability: ability of something to burn
Reactivity: ability of something to react with another material (sodium is very reactive in water, noble gasses generally do not react with other materials)
Toxicity: ability of a material to act as a poison (arsenic)
Corrosiveness: ability of a material to "eat through" another material (strong acid or strong base)

Chemical Change: process of creative new compounds
Burning wood, rusting iron, chemical reactions, formation of water from its component gasses

Names Period Date Teacher

Properties of Matter Lab

Problem: What do various materials have in common with each other?

Research: At each station, you will use your powers of observation.

Hypothesis: The items can be described in simple terms.

Experiment: You will have 4 minutes to examine the objects at each station. You must describe each object five different ways in the data tables provided in this lab packet. Tools at each station, such as rulers, balances, etc., are not to be described.

Materials: Each station will have a set of five different items, labeled A, B, C, D, and E. You do not need to know the names of these items, only the letter.

Analysis: For each station, you must write similarities and differences among the objects.

Conclusion:

- Why is it important to be able to describe different objects?
- What are some categories you used to describe the objects?
- What are some other possible categories you might have been able to use for these objects or other objects?
- Complete the table on the back based on your data. Do not use the names of the objects when completing the table. Use only the letters on the objects.

Reflection:

- What were your favorite parts of this lab?
- What were your least favorites parts of this lab?
- What would you change about this lab to make it better?
- How well did you work within your group?
- What could you have done differently to make working in your group even better?

For each station, write a yes/no question that could separate the five objects into two distinct groups. Your questions must be grammatically correct. Write the letter of each object in the Yes/No box, depending on how the object answers the question. See the sample below.

SAMPLE: Materials: A = circle, B = square, C = triangle, D = oval, E = rectangle

Sample Question: Does the object have four sides?

Yes	No
B, E	A, C, D

Station 1 Question:

Yes	No

Station 2 Question:

Yes	No

Station 3 Question:

Yes	No

Station 4 Question:

Yes	No

STATION 1	A	B	C	D	E
Describe five things about each object at the station.					

Similarities		Differences	
List 1 to 3 similarities between the 5 objects at this station.		List 3 to 5 differences among the objects at this station.	

STATION 2	A	B	C	D	E
Describe five things about each object at the station.					

Similarities		Differences	
List 1 to 3 similarities between the 5 objects at this station.		List 3 to 5 differences among the objects at this station.	

STATION 3	A	B	C	D	E
Describe five things about each object at the station.					

Similarities		Differences	
List 1 to 3 similarities between the 5 objects at this station.		List 3 to 5 differences among the objects at this station.	

STATION 4	A	B	C	D	E
Describe five things about each object at the station.					

Similarities		Differences	
List 1 to 3 similarities between the 5 objects at this station.		List 3 to 5 differences among the objects at this station.	

Mixture Terminology Food Lab – Teacher Notes

Like the *Properties of Matter Lab*, this relies on students using their powers of comparing and contrasting to solve each station. This time, scientific terminology will be present at each station and students are challenged to define the terms. This lab takes a fair amount of setup to label everything.

<u>**Time**</u>: One class period

<u>**Materials per station**</u>

Each material is in its own beaker (total: 21) and must be covered to help prevent students from sampling the food. I generally set up two sets of stations so I can have smaller student groups.

Station	Container 1	Container 2	Container 3
1	Chocolate powder [solute]	Milk * [solvent]	Chocolate milk * [solution]
	Sugar [solute]	Unsweetened iced tea ** [solvent]	Sweetened iced tea ** [solution]
2	Plain water / spoon [unsaturated]	Water with excess salt / spoon [saturated]	Small amount of salt
3	Beef stew [heterogeneous]	Tomato soup [homogeneous]	
	Lucky Charms cereal (or similar) [heterogeneous]	Cheerios cereal [or similar] [homogeneous]	
4	Raspberry Kool-Aid ***	Blueberry Kool-Aid ***	Mixture of the two [miscible]
	Water	Vegetable oil	Mixture of the two [immiscible]
5	Mustard [colloid]	Italian salad dressing [suspension]	

* I generally use glue with water in place of milk so it won't spoil
** You can use the same tea for both; they're not tasting it
*** Food coloring and water is all you need to make red, blue, and purple solutions

<u>**To the class**</u>

Today you will use your powers of observation to define various scientific terms. By comparing and contrasting, you should be able to determine what the terms means. [This is like the properties of matter lab we did.] At station 2, add a small pinch of salt to each beaker and stir. At stations 1, 3, 4, and 5, make comparisons. How do the samples relate to each other? For lab safety today, absolutely DO NOT eat or drink any food item you see. Some of them are not what they seem. The milk, for example, is glue and water. The cereals, I've had for years. Do not unseal the beakers.

Lab Guidance

Students will be confused about what to do. Walk around and give some guidance so they understand.

Station 1: If you're going to make chocolate milk, how would you do it? What goes where?

Station 2: When you added the salt, did it dissolve or just sink to the bottom?

Station 3: Think about what exactly would be on your spoon each time you scoop out a sample.

Station 4: What's different about these two mixtures?

Station 5: If you were going to use these, would you have to do anything special first?

Discussion

I generally put the terms on the board, then check student answers. I have those who are correct (or nearly so) put their answers on the board for one station. We then go through the definitions and students make updates on their lab sheets.

Station 1

Solute: a material that gets dissolves (chocolate powder, sugar), often a solid (but doesn't have to be)

Solvent: a material that does the dissolving (milk, unsweetened iced tea), often a liquid

Solution: the combination of solute and solvent (chocolate milk, sweetened iced tea) [Note: true solutions are supposed to be clear, so the chocolate milk is just representative.]

Station 2

Unsaturated: a solution that is NOT full and CAN dissolve more solute

Saturated: a solution that is FULL and CANNOT dissolve more solute

__Supersaturated__: a solution with MORE dissolved than normal [this was not on their lab sheet]

Station 3

Heterogeneous: a mixture that is different throughout (beef stew, Lucky Charms), every scoop of a spoon will have different ingredients

Homogeneous: a mixture that is the same throughout (tomato soup, Cheerios)

Station 4

Miscible: two liquids that are able to add together (Kool-Aids)

Immiscible: two liquids that cannot add together (oil and water)

Station 5

Colloid: a mixture that DOES NOT need to be shaken (or stirred) before use

Suspension: a mixture that DOES need to be shaken before use

Food lab labels

Solute (A) Sugar	Solvent (A) Unsweetened Iced Tea	Solution (A) Sweetened Iced Tea	Unsaturated
Solute (B) Chocolate Powder	Solvent (B) Milk	Solution (B) Chocolate Milk	Saturated
Homogeneous (A) Tomato Soup	Heterogeneous (A) Beef Stew	Homogeneous (B) Cheerios	Heterogeneous (B) Lucky Charms
Liquid (A) Raspberry Kool-Aid	Liquid (A) Blueberry Kool-Aid	Miscible (A) Raspberry and Blueberry Kool-Aid	Salt Add one pinch of salt to each beaker and watch
Liquid (B) Water	Liquid (B) Vegetable Oil	Immiscible (B) Water and Vegetable Oil	Colloid Mustard
			Suspension Italian Salad Dressing

Names Period Date Teacher

Mixture Terminology Food Lab

Compare and contrast at each station. Use your skills to develop a working definition for each of the scientific terms below. The definition needs to be GENERAL so it can be applied to more things than what you see at the station. Do not list the items you see at the station.

Station 1

solute	
solvent	
solution	

Station 2

saturated	
unsaturated	

Station 3

heterogeneous	
homogeneous	

Station 4

miscible	
immiscible	

Station 5

colloid	
suspension	

Mixture Separation Techniques – Teacher Notes

In this activity, students will separate a mixture of salt, sand, iron filings, and pebbles. You can run this lab in two ways. First, you can give specific instructions on the order of separations. In this case, discuss the lab and talk through the desired procedures given on the next page. Second, you can allow students to devise their own plans and carry them out in that order. Also, to separate sand and water, hot plates will be necessary, therefore goggles are a must, as is a reminder of lab safety. If you are uncomfortable with using hot plates, you can simply skip the salt and water separation.

Time: One class period

Materials per group

- A small cup with the mixture to separate (pebbles, iron filings, sand, salt)
- Access to water (50 mL is fine)
- Wire screen
- Magnet (preferably in a sealed bag)
- Coffee filter
- 4 beakers: 100mL or 250mL
- Goggles per person (if using the hot plate)

Materials per class

- Hot plates (two are sufficient for six to eight groups)
- Beaker tongs (for teacher use only)
- Watch glass per beaker to cover the beaker when on the hot plate

To the class

Today you will be challenged to separate a mixture into its component parts. You will have sand, iron filings, salt, and pebbles to separate. You need to come up with a plan for how to separate these materials. You will have several materials you can use, and these should guide you as you make your plan. You will also have access to water. We will be using the hot plates today, so goggles must be worn at all times. Remember, once a beaker is placed on a hot plate, you CANNOT and MUST NOT touch it again. I will remove the beakers when they are finished. You must not touch the beaker even after it's off the hot plate. It will still be hot for a while. For clean up, you will discard the wet filter paper and rinse out the wet beakers. [You can have them pour the separated materials into your stock mixture, but I usually don't do this in front of them, as it makes the lab seem pointless.]

Separation Techniques (in order of the lab)
Sieving: using a screen to separate particles of different sizes (pebbles get left behind)
Magnetism: using a magnet to draw certain metals (iron filings are pulled out)
Filtration: add water to the salt and sand, then pour through the filter (sand is left behind)
Boiling: heating up the liquid until it vaporizes, leaving behind the solute (salt remains in beaker)
* **Distillation**: separating different liquids based on their boiling points
* **Chromatography**: separate particles by passing them through a medium, separates by polarity

* these techniques were not performed during this experiment

For added guidance, a placement is included on the next page. This is not required but may be helpful.

Beaker with pebbles 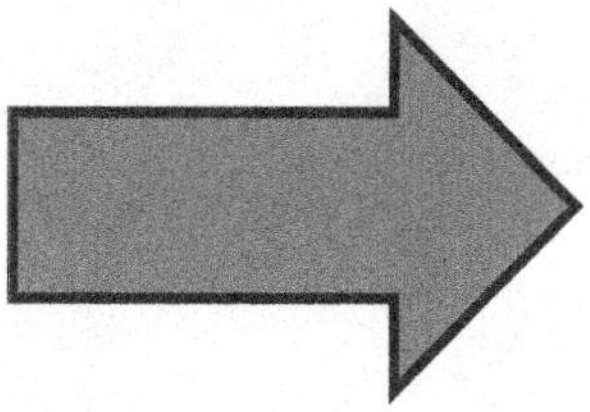

Beaker with iron filings 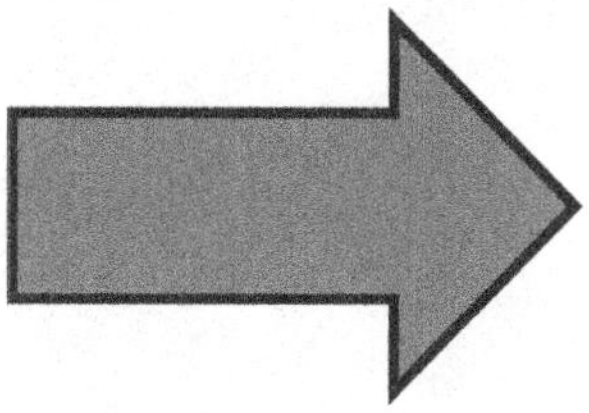

Beaker with coffee filter and sand 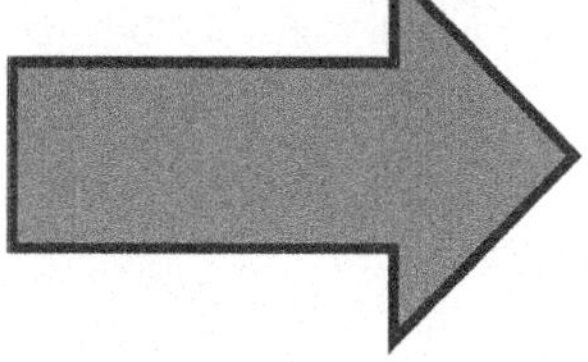

Names Period Date Teacher

Mixture Separation Techniques Lab

Problem: You will be given a dry mixture and you need to separate it into all its individual parts.

Research: Use your knowledge of the phases of matter and everyday materials to help you. For each situation below, discuss and describe how you could possibly separate the items.

Items to Separate	Possible Method(s) to Perform the Separation (give some description)
Sand and water	
Salt and sand	
Iron filings and sand	
Rocks and sand	
Rocks and iron filings	
Water and salt	

Materials: Give a list of materials that might help you to complete the separations you described above.

Challenge: You will be given a dry mixture that contains (1) iron filings, (2) rocks (or pebbles), (3) salt, and (4) sand. Your job is to separate all of these into separate beakers by the time the period is up. Please note that these items are listed alphabetically and are not necessarily in a helpful order. You will have access to some empty beakers, a graduated cylinder, a hot plate, 50 mL water, a coffee filter, a wire screen, and a magnet. You may not use the hot plate without permission.

Hypothesis: Read the challenge again carefully. Come up with a plan that will allow you to separate all four solids into individual containers without contaminating them. If you mess up, you have to try to figure out a way to fix it. So plan carefully and think it through with caution. In the spaces below, write out the order in which you plan to do the separation. Don't just use the order given to you in the Challenge unless you really think that's the way to go! Fill in the left-side column first. As you make your way through the lab, you may realize you have to make changes. Once you're finished, write the actual steps in order on the right.

Steps	Your Preliminary Guess	Your Actual Process
First, separate out the…		
Second, separate out the…		
Third, separate out the…		
Last, separate out the…		

** DO NOT USE THE HOT PLATE WITHOUT PERMISSION OR SAFETY GOGGLES. **

Experiment: Once you have a plan, show your teacher. Then get to work. Make notes along the way. Briefly explain what you are doing at each step and then what you managed to separate out. Also, briefly explain if what you did worked out as you expected and what you might do differently. An example is shown for reference…

Example:

What you did…	What you separated out…	How you could do it better…
Step 1. We added water to the dry mixture right away. We thought if the rocks were wet, they'd slip out easily.	We were able to pull the rocks out by hand but they were coated in wet sand and salt. Taking them out one at a time seemed like a waste of effort.	The whole thing is a mess! We would have been better keeping the mixture dry until we had no other options. Also, there's gotta be a better way to get the rocks out without picking them out one at a time.

Your Notes:

Analysis: List the materials were you able to successfully separate: ______________________________

Conclusion: (3-6 sentences) There is a way to separate out all the materials into separate beakers without contaminating them along the way. Consider the steps you took and describe any improvements you could make to the process you used. What changes did you make during the experiment that you hadn't planned on? Also, be sure to fill in "Your Actual Process" on the front side.

Fried Ice – Teacher Notes

Graphing is an important skill, and we can use data from melting ice and boiling the water to create a graph that relates to the heating curve for water in its various phases. This is usually the first graphing assignment I give in class, so I have the students graph it in class so I can observe how they do it. I generally have all students record data on their own lab sheets for this activity rather than having one sheet per group like normal.

Lab Safety

For this lab, you will need to use hot plates. Therefore, lab safety must be made a serious concern. Students are not to touch the beaker or the hot plate surface once the hot plates are turned on. Goggles must also be worn at all times. While the students are running the lab, pay attention to the clock. You may need to increase the heat setting of the hot plates to ensure the samples will boil before time is up.

Time: One class period (or two if you want to take time showing them how to make a graph properly)

Materials per group

- Hot plate
- Safety goggles
- 250 mL beaker
- 100 mL water
- Ice cubes (or 1 paper bathroom cup ice cube)
- Thermometer
- If available, ring stand and thermometer clamp

To the class

Today we're using the hot plates, so safety is of utmost importance. You must remember not to touch the surface of the hot plate or the beaker once it is set on the hot plate. These will become extremely hot even though they won't look like it. You will start with a beaker of ice water and you'll record its temperature for a couple of minutes, stirring before you take the temperature each time. After that, you will turn on the hot plate halfway up and start a stopwatch. Every minute, stir and record the temperature again. Repeat this until the end of the lab. For cleanup, simply turn off the hot plate and set the thermometer on the counter. Do not touch the beaker. I will remove it. We will graph the data tomorrow [or for homework].

Discussion

It takes heat to change phases from solid to liquid to gas. We added heat using the hot plate and saw the ice turn to water, which turned to steam. The first few readings should be pretty close in temperature, as should be last few readings. This depends if the sample was able to boil in time.

Draw a rough sketch on the board, as shown in the graph on the next page. Make a note of the beginning few pieces of data and the last few and show how they are pretty close together in temperature. Draw a horizontal line of best fit across these two areas, as shown. These represent the **heat of fusion** (HF) between ice and water and the **heat of vaporization** (HV) between water and steam. Label these. From the inner points of these lines, draw one continuous line to connect them, representing the liquid phase. Then extend the graph downward on the left and upward on the right to represent the solid and gas phases, respectively.

From here you can discuss the six major phase changes, which students can write onto their lab sheets.

Heating Curve Information

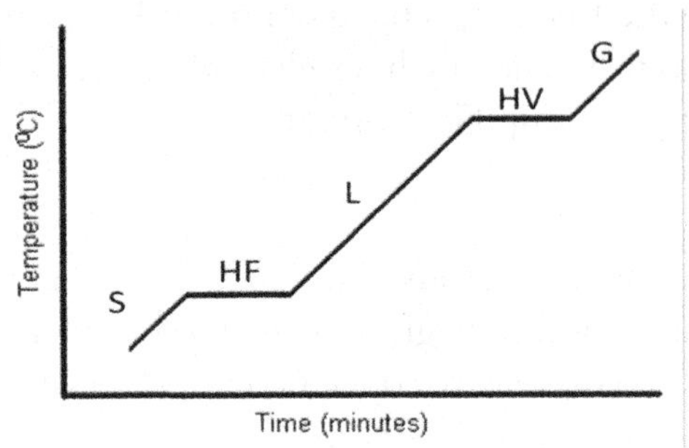

Phase Changes		
Phase Change	**Phases Involved**	**Example**
Melting	S → L	ice → water
Freezing	L → S	water → ice
Boiling	L → G	water → steam
Condensing	G → L	steam → water
Sublimation	S → G	dry ice → CO_2 gas
Deposition	G → S	CO_2 gas → dry ice

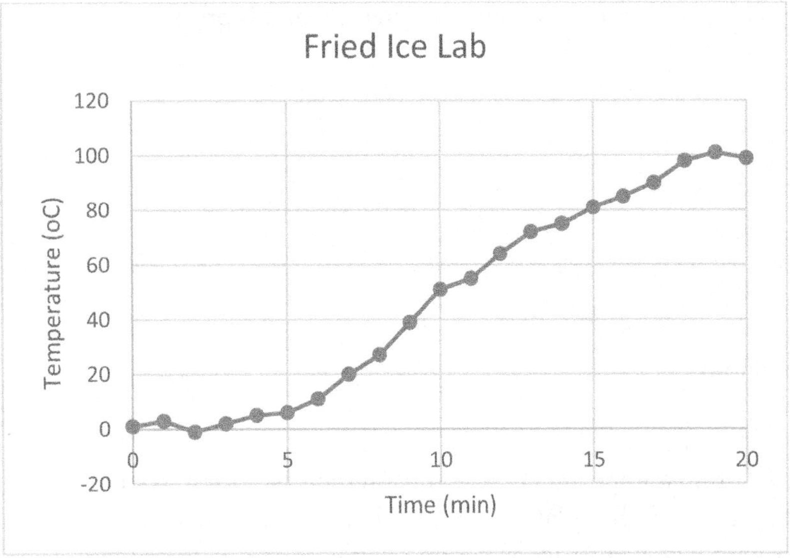

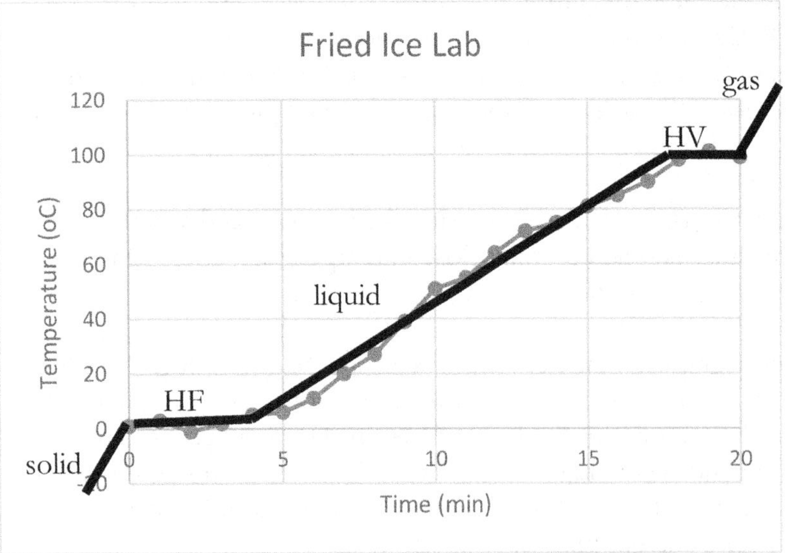

Name Period Date Teacher

Fried Ice Lab

In this lab, you will melt a sample of ice down until it boils (hopefully). You must adhere to proper lab safety rules for this lab or you will be ejected from the classroom.

Lab Safety

1) You will be using a hot plate today. It is imperative that you DO NOT PLACE ANYTHING onto the white plate surface except the glass beaker.
2) DO NOT place your hands, pens, pencils, paper, or any other material on the top of the hot plate.
3) You must also wear safety goggles AT ALL TIMES when you are near a hot plate. Write your initials in the box if you agree to follow these rules.
4) If you do not agree, then you cannot do this lab, and you be given another assignment.

I agree to follow lab safety procedures.

Procedure

1) Put on your safety goggles.
2) Add 100 mL of water to a 250 mL beaker. (It does not have to be exact.)
3) Tear the cup off your block of ice. Place the ice into the beaker with water. Discard the paper cup.
4) Place the beaker with the water and ice on top of the hot plate.
5) Set the thermometer into the beaker and stir. Record the initial temperature as Time 0 min (zero).
   a) Repeat for two more minutes.
6) Turn the heat up "halfway".
7) Start the stopwatch.
8) Wait one minute.
9) Gently stir the water/ice with the thermometer for a few seconds.
10) Read and record the temperature to the nearest whole number.
11) Repeat steps 7 through 10 until time is up.
12) When the lab is finished, set the thermometer on the counter, turn off the hot plate, and walk away.

Time (min)	Temperature (°C)
0	
1	
2	
3	
4	
5	
6	

Time (min)	Temperature (°C)
7	
8	
9	
10	
11	
12	
13	

Time (min)	Temperature (°C)
14	
15	
16	
17	
18	
19	
20	

Graph Notes

- Graphs are set up in a specific way:
- The INDEPENDENT VARIABLE always goes on the X-AXIS (horizontal).
- The DEPENDENT VARIABLE always goes on the Y-AXIS (vertical).
- Each axis must be counted in a pattern, and not simply be a copy of the data.
- Each axis must be labeled with the variable name and its measured unit.
- For a line graph, plot each point as a visible dot. Connect each dot from left to right.
- Include an appropriate, descriptive title that does NOT repeat the variable names.

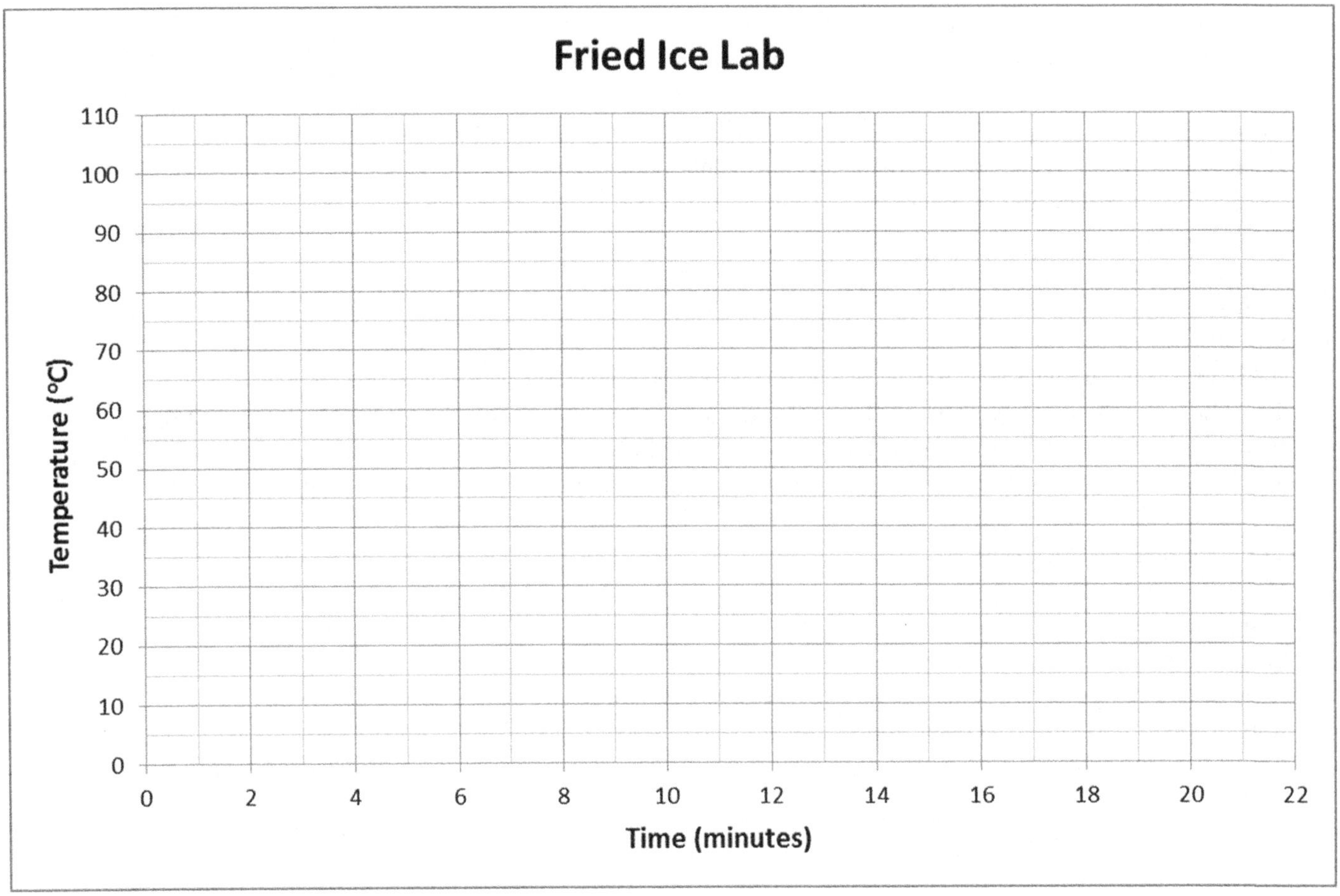

Heating Curve Information	Phase Changes		
	Phase Change	**Phases Involved**	**Example**

Chapter 2

Density

Density is a property of matter. It belongs to the material itself. It's a measure of how close molecules are in a substance. When it comes to density, sample size does not matter. The density won't change if you have a block of wood or if you cut that block in half. It's the same material, therefore the density does not change. This is a common misconception. Another common misconception is that density is the same as mass. It's not. It's the amount of mass in a given amount of space (volume).

Density is calculated by dividing the mass of the sample by its size using this formula.

$$density = \frac{mass}{volume}$$

The units for density will change depending on the units used in the calculation. For instance, you could have g/mL (which is often measured in a school setting), g/cm^3 (also common for a school setting), kg/L, and so on.

Density can be used to determine whether an object will float in a liquid. If the density of the object is less than the density of the liquid, then the object will float. Most students know that the density of water is 1.00 g/mL. However, they mistakenly think that as long as the density of an object is less than 1, then the object will float. That's only true if the liquid is water. For example, vegetable oil has a density of 0.86 g/mL. A block that has a density of 0.92 g/mL would sink in oil but float in water.

The following labs explore aspects of density. We begin with a research task for the students.

Name Period Date Teacher

Density Research

Use the internet to find the answers to the questions that follow. Be sure to keep a record of the precise sources you use. Include these at the end of the document.

1. What is mass? What are some units used to measure mass?

2. How are mass and weight different?

3. What is volume? What are some units used to measure volume?

4. What is density? What are some units used to measure density?

5. What formula is used to calculate the density of an object?

6. Provide an example of a density calculation.

7. You have an object with a density of 3.86 g/mL and a volume of 23.1 mL. What is the object's mass?

8. A block has a density of 1.6 g/cm^3. If you cut the block in half, its mass reduces by half and so does its volume. What is the density of one of the halves of the block?

9. How does density determine whether an object will float or sink?

10. You have three objects with different densities.

Object	Density
A	0.8 g/mL
B	0.25 g/mL
C	1.25 g/mL

a. If these were liquids, which one would float on top?

b. If the density of water is 1.0 g/mL, which object(s) will float?

c. If the density of a liquid is 0.6 g/mL, which object(s) will float?

Sources

Water Measurement Lab – Teacher Notes

This lab could also be called the accuracy versus precision lab. In this, we're calculating density using various pieces of equipment (graduated cylinders and beakers) to see which one(s) produce consistent results of water's density, 1.00 g/mL.

<u>Time</u>: One class period

<u>Materials per group</u>

- Balance
- 10 mL graduated cylinder
- 50 mL graduated cylinder
- 100 mL graduated cylinder
- 100 mL beaker
- 250 mL beaker
- water

<u>Background information</u>

Accuracy is when a measurement is close to a given, accepted value, such as the density of water at 1.00 g/mL.
Precision is when various measurements are close to themselves.
A common analogy is a dartboard. The closer the darts are to the bull's-eye, the more accurate they are. The closer the darts are to each other (say they all landed in the 3-point zone), the more precise they are.

<u>Notes</u>

My students always forget to record the precise volume when taking measurements, so I make sure I help each group with their first set of measurements. They need to fill a beaker (for instance) one-third full of water. For a 100 mL beaker, this should be 33 mL, but there is no marking on the beaker for that, so they end up estimating. Once the water is in the beaker, they should look again to see where the water is. Is it really 33 mL or something else? Whatever it is, record it.

The results for the lab can be messy at times. There is a lot of inaccuracy of the measuring tools, so that will cause data to be skewed, which is partly the point of the lab.

When plotting the data, each tool will only have three data points and they won't all stretch across the graph. This may cause some confusion. If students have densities greater than 1.30 g/mL, they made some error in their measurements and should try it again if there's time. If not, have them plot the value on the 1.30 line.

<u>To the class</u>

We're going to test the accuracy and precision of different measuring tools to see which one provides the most accurate and precise data. [Recap accuracy and precision as needed.] For this lab, you need to read measurements as carefully as possible, down to decimal places if the equipment allows for it. You're going to start by getting the mass of a graduated cylinder to the nearest tenth [or hundredth if your equipment can handle that]. Add water so it's about one-third full. It doesn't have to be exact. Then – and this is a part that people forget to do – you have to read the actual volume in the graduated cylinder and record it in the data table. [Show them where.]

Obtain the mass of the graduated cylinder and the water and record that too. The rest is calculations. [Discuss if you want them to do it at each step or save that for individual assistance.] Next, add more water so it's up about halfway, record its volume and its mass. Finally, add water up to the top line. Not the top of the graduated cylinder – just to the top line. Record its volume, then get the mass. Move on to test another piece of equipment.

When plotting your graph, use a different color for each container. You will only have three points per container and the lines may be small when you connect them. Do your best.

Discussion

Results will likely vary, but the 10 mL graduated cylinder should have had the most accurate value for density, and likely the most precise. If, for instance, someone says their data showed that the 100 mL graduated cylinder was the most accurate, ask why. (Most probably, it was the easiest to read the lines.)

The 250 mL beaker should have come out as the least accurate and precise. This is because the divisions are large values; depending on the type of beaker you have, these could be gradations of 50 mL, leaving a lot of room for error.

Regardless of the type of equipment used, there is always error in reading the amount of liquid inside. Use the most accurate measuring device for the type of lab you're conducting. If you need volumes close to 100 mL, use a 100 mL graduated cylinder. If you need smaller volumes, use smaller cylinders.

Names Period Date Teacher

Water Measurement Lab

Purpose: In this lab you will measure the density of water using various pieces of lab equipment. Your goal is to determine which equipment has the best and worst accuracy and precision.

Procedure

1) Obtain a container, then measure and record its tare weight while empty (and preferably dry)
2) Fill the container one-third full of water and **record the actual volume** to the best precision of the container
3) Obtain the mass of the container with water, then subtract the tare weight to get the mass of water
4) Add more water until the container is half full, record the actual volume, then find the mass of water
5) Finally, fill the container to the highest measurement printed on it, record the volume, then obtain the mass of water
6) Repeat for different containers as time permits
7) Calculate the density for each sample by taking the water mass and dividing it by the actual volume; keep two decimal places for each calculation

Data Table

Container	Container Tare Weight (g) [A]	Mass with Water (g) [B]	Water Mass (g) [B-A]	Estimated Volume (amt)	Actual Volume (mL) [C]	Density (g/mL) [(B-A) / C]
50 mL graduated cylinder				One-third		
				One-half		
				Top line		
100 mL graduated cylinder				One-third		
				One-half		
				Top line		
50 mL beaker				One-third		
				One-half		
				Top line		
100 mL beaker				One-third		
				One-half		
				Top line		
250 mL beaker				One-third		
				One-half		
				Top line		

Analysis

Create a graph for the densities calculated in the data table. You will need a separate line for each container, preferably drawn in a unique color.

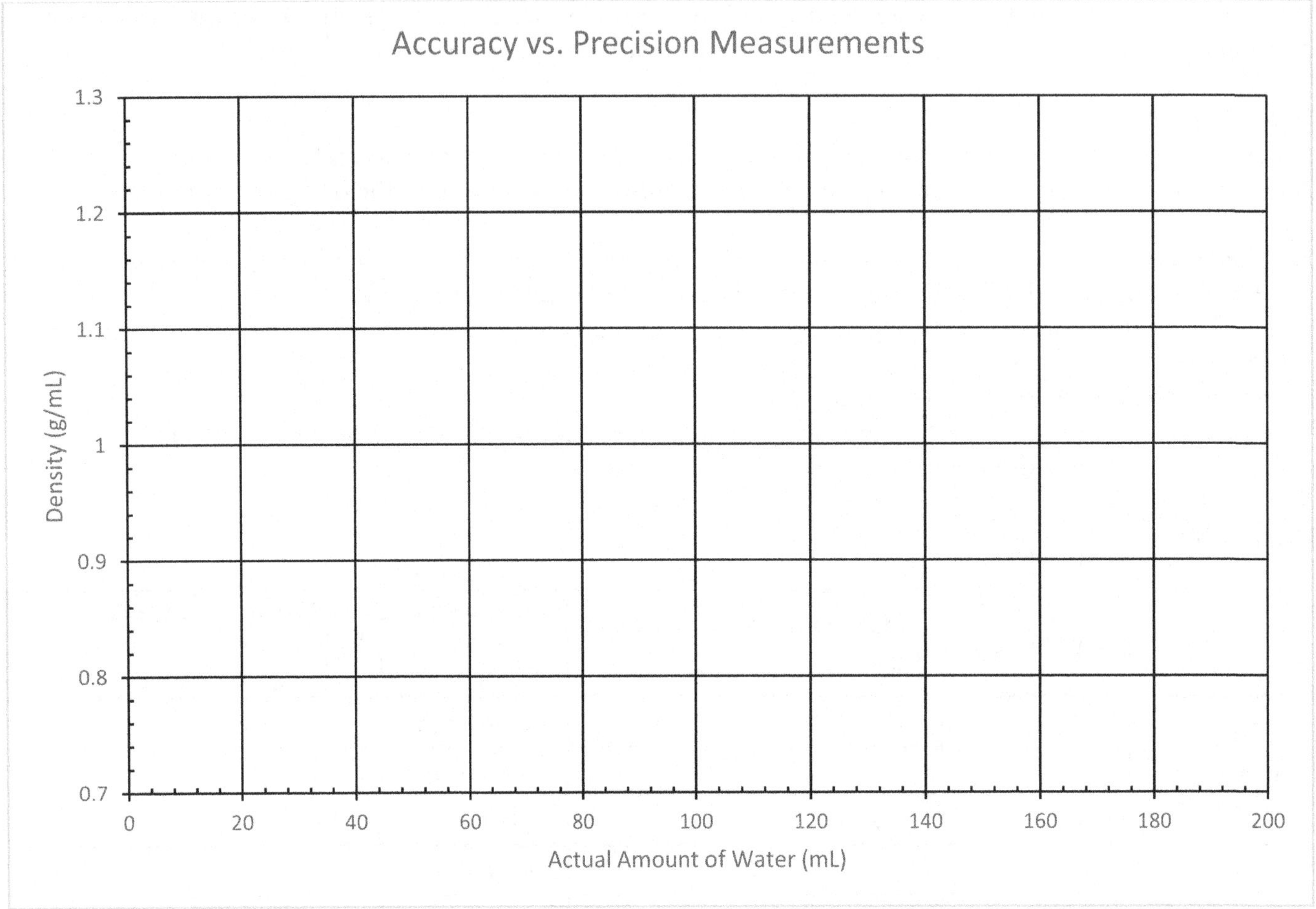

Which container had the best accuracy? (line closest to 1.00 g/mL)	Which container had the least accuracy? (line furthest from 1.00 g/mL)

Which container had the best precision? (flattest horizontal line)	Which container had the least precision? (most jagged or steepest line)

Saltwater Density Lab – Teacher Notes

In this experiment, students will calculate the density of water and saltwater after predicting what will happen to the density.

Time: One class period

Materials per group

- Two 250 mL beakers (one remains dry)
- Approximately 125 mL water
- 100 mL graduated cylinder
- 75 grams of salt
- Stirring stick (or spoon)
- Electronic balance *

* If you're using a triple beam balance instead of an electronic balance, you'll need to show the students how to obtain the masses by use of subtraction. For instance, get the mass of the graduated cylinder empty, fill it with 90-100 mL water, then get its mass, then subtract the two values to get the mass of the water. The same process will be needed for obtaining the mass of the salt.

Notes

Students tend to forget to record the volume. Some also try to pour all 125 mL into the graduated cylinder instead of stopping between 90 and 100 mL. Additionally, students tend to hit the zero button every time they place an object on the balance. This will ruin the tare weight of the graduated cylinder and it will need to be measured again. Students also want to add 50 and 75 grams of salt to the solution in the wet beaker. This is incorrect. They're only adding 25 g more. It's important to make this clear.

In the cleanup instructions, I say not to clean the dry beaker. This is only necessary if you have another class repeating the lab the same day.

To the class

> Today we're going to measure the density of water and saltwater. You will be making use of the balances and some glassware. You need to follow the procedure carefully and go step by step or you will make mistakes. [Read through the procedure and show them how to get the tare weight.] Remember, after you get the tare weight, DO NOT touch the zero button again until after you record the new mass of water.

Discussion

Not much of a surprise, but students should see that the density increases as more salt is added.

Names Period Date Teacher

Saltwater Density Lab

Purpose. In this lab, you're going to measure the density of freshwater and saltwater to make a comparison between them. Please take care not to spill water on your table.

Problem: Restate the purpose in the form of a question.

Research: What's something you already know about salt and water?

Background information: Density is a measure of how tightly packed molecules are in a substance. It can be used comparatively to determine if one substance will float in another substance. The material with the lower density will float on top. Density can be determined with the following formula:

$$density = \frac{mass}{volume}$$

Getting the tare weight: Turn on the balance. Use the mode button if necessary to ensure the balance is set to measure in grams. Place the container on the balance. Press the "on" button. If you're using a triple beam balance, set it to zero, then place the beaker on it and find its mass. Then move the sliders up by 25 grams. You will add salt until the scale balances again after this.

Hypothesis: What will happen to density as more salt is added?

Experiment: Follow the procedure carefully. The procedure is on the back.

Materials: Dry beaker. wet beaker, salt, balance, water, stirring stick

Constants: What are three constants you should control to make this a valid experiment?

Data Type: Will you be collecting qualitative data or quantitative data? ___________________________

Procedure:

1. In your **wet beaker**, add approximately 125 mL water. It does not need to be exact.
2. Prepare for measurements by setting the tare weight of the graduated cylinder. It's okay if it's wet.
3. Remove the graduated cylinder from the balance temporarily.
4. Add between 90 mL and 100 mL water to the graduated cylinder. Record the exact volume used.
5. Place the graduated cylinder on the balance again and record the mass without resetting the scale.
6. Calculate the density and record to the nearest hundredth.
7. Empty the graduated cylinder into the **wet beaker**. You do not need to dry it.
8. Tare the **dry beaker**.
9. Add 25 grams of sodium chloride (salt) to the **dry beaker**.
10. Pour the salt into the **wet beaker**. Make sure the dry beaker remains dry.
11. Stir for a couple minutes until most of the salt dissolves. It may not all dissolve.
12. Repeat steps 2 through 5 to determine the density of the saltwater.
13. Measure out another 25 grams of salt and repeat steps 8 through 11.
14. If time permits, carry out another trial by adding another 25 grams of salt.
15. Clean up. Do not wash the **dry beaker**. Leave it empty and dry.

Data Table:

Amount of Salt	Water Mass (g)	Volume (mL) ± 1 mL	Density ($^{g}/_{mL}$) ± 0.01 $^{g}/_{mL}$
0 grams			
25 grams			
50 grams total			
75 grams total			

Analysis:

What was the trend for the data?

Conclusion:

Was your hypothesis correct? Why?

What are some possible errors for this experiment?

How can the experiment be improved?

Water Column Lab – Teacher Notes

Students will create a four-tier water column with different colors separated out. This depends upon the densities of the different solutions, where the lower densities will sit on top of the higher densities. This lab can be tricky to do correctly. It often comes down to how well the students prepare their solutions and how carefully they add each water sample to the test tube.

Time: One class period

Materials per group

- 1 test tube
- Test tube rack
- Food coloring – 4 colors
- Four 250 mL beakers
- Four plastic droppers
- Four stirring sticks
- 60-100 g salt
- 1 plastic teaspoon

Note

When students add the second, third, and fourth colors to the test tube, they need to lower the dropper tip as close to the previous liquid layer as they can *without touching* it. I also suggest tipping the test tube slightly so the new water doesn't splash. They should add the water slowly, not squirt it out.

If no one is having success, consider doubling the amount of salt in each cup. It might be worth trying this out on your own first to see if you need to make this modification.

To the class

In today's lab, we're going to make what's called a density column using water with different densities. [If you recall from our previous lab, when we added salt to water, the density changed. We're going to follow a similar procedure here.] The lab can be tricky to get right. You have to make your solutions carefully. Also, when you add your solutions to the test tube, you have to be cautious. If you drop the solution in from too high up or splatter it, it will mix with the other layers and it won't work. When done correctly, you should see all four colors in the test tube.

Discussion

- Which solution had the highest density? [red with the most salt]
- How do you know? [it sat at the bottom]
- If you created a purple solution with 1.5 teaspoons of salt, where would it float? [between blue and green] (Note: if you doubled the salt, increase this value to 3 here and on the lab sheet.]
- Did you have time to add the colors in a different order? What happened? [they mixed]

Names Period Date Teacher

Water Column Lab

Purpose: In this lab, you're going to prepare solutions with different salinities, then you will combine them together and observe the results.

Problem: Restate the problem in the form of a question.

Research: What is something you already know about this topic?

Background information: Density is a measure of how tightly packed molecules are in a substance. It can be used comparatively to determine if one substance will float in another substance. The material with the lower density will float on top.

Hypothesis: What do you think will happen in this lab?

Experiment: Read and follow the procedure carefully.

Materials: Food coloring, four cups, four droppers, salt, spoon, stirring sticks, test tube, flask or test tube rack

Procedure:

1) Fill each beaker one-third full of water.
2) Add 1-2 drops of food coloring to each beaker.
3) Keep the teaspoon dry at all times. Use stirring sticks to stir the solutions below.
4) Dissolve 1 teaspoon of salt in the blue beaker.
5) Dissolve 2 teaspoons of salt in the green beaker.
6) Dissolve 3 teaspoons of salt in the red beaker.
7) Do not add salt to the yellow beaker.
8) Use a separate dropper for each color.
9) Squeeze the dropper and draw up the red solution.
10) Place the dropper inside the test tube and slowly squeeze out the liquid near the bottom surface.
    a) When other liquids are in the test tube, do not touch them with the tip of the dropper.
11) Repeat steps 8 through 10 with the green, blue, then yellow solutions, in that order.
12) Sketch the results on the back.
13) The solutions should have separated. If they didn't, and time permits, try again.
14) If time permits, clean out the test tube and droppers and try adding colors in a different order.
15) Clean up your station. Rinse everything out with water and return it to your station. No soap. Clean out and save the droppers (fill and squirt them with water).

Analysis: In the diagram, draw what you saw in your test tube. Use colored pencils if available. If you did more than one trial, color those in too.

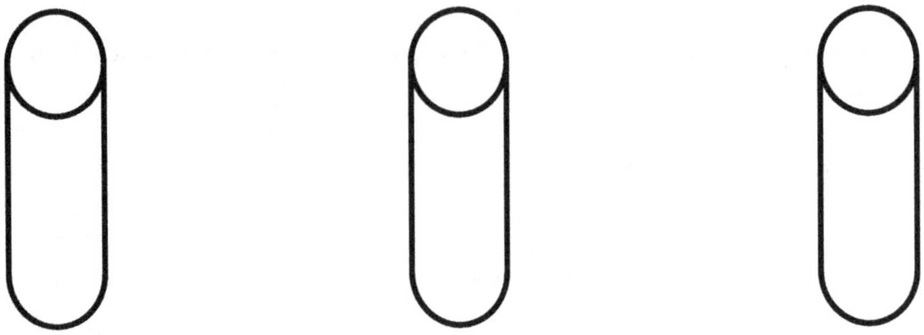

Conclusion:

The solutions should have separated. Did yours? ______

Why should the solutions separate?

What would happen if the solutions were added out of order?

If you created a purple solution with 1.5 teaspoons of salt, between which two layers should it float?

Density of Unknown Metals – Teacher Notes

Because density is a property of matter, it can be used to help identify materials. Copper, for instance, has a density of 8.96 g/mL at STP regardless of sample size or shape. Students will use this property of matter to identify various unknown metals.

Time: One class period

Materials per group

- Balance
- 100 mL graduated cylinder (or 50 mL if the cylinders fit easily)
- Ruler

Materials for the class

- Various metal samples to test of differing size and/or shape
- Copper, aluminum, and iron work well

Notes

I use the graduated cylinder to test cylinders and irregular-shaped objects using water displacement. I have students measure the volume with the ruler for cubes and rectangles. If the cubes fit in the graduated cylinder, I sometimes have them also obtain the volume using that method as a means for comparison. I will often have cubes and cylinders of each material for comparison. If you don't have cubes and cylinders of these metals lying around, you can use pieces of copper pipe, iron nails, and sheets of aluminum foil folded into a thick enough cube that students can measure with the ruler.

To the class

Today we're going to use the property of density to identify unknown metals. Density is a physical property of matter, so it "belongs" to the material. Sample size and shape shouldn't matter, and we're going to test that out. For cubes and rectangles, you'll be measuring length, width, and height using a ruler. You have to be careful to line up with the zero line and measure accurately to the nearest tenth. For cylinders, we'll use water displacement by seeing how much the water level changes when the cylinder is inserted into the graduated cylinder.

Discussion

The densities measured should be relatively close to those in the chart. However, there will be differences. For one, the rulers are not precise enough considering the cubes are likely to be small. The graduated cylinder will also lack precision, moreso if it's larger. This produces error in the volume measurements, and since the samples are small, this translates into larger errors. You could extend this lab by using different graduated cylinders and utilizing calipers for the other volume measurements.

Names Period Date Teacher

Density of Unknown Metals Lab

Purpose: To determine the identities of unknown metals based on their densities.

Background: Density is a property of matter. This means that all samples of a particular material will have the same density under normal conditions. Let's say you have a block of wood with a density of 0.6 g/mL. You cut the block half. The density of each half will still be 0.6 g/mL. It's not 0.3 g/mL! Why? Because density relates to the material, not the sample size or shape. Density can be found by the following formula.

$$density = \frac{mass}{volume}$$

Because density stays the same for a given material, we can use density to help identify unknown materials, regardless of the sample's shape or size. Here are the densities of some metals.

Name of Metal	Density (g/mL)	Name of Metal	Density (g/mL)
Aluminum	2.70	Magnesium	1.74
Copper	8.96	Nickel	8.91
Gold	19.32	Silver	10.50
Iron	7.87	Tin	7.287
Lead	11.35	Zinc	7.134

Procedure

1) Use a balance to obtain the mass of an unknown sample of metal. Record this in the data table.
2) For cylinders:
   a) Fill a graduated cylinder about halfway with water. Record the exact volume in the data table.
   b) Tilt the graduated cylinder and slide the metal inside without splashing.
   c) Record the new volume in the data table.
   d) Subtract the two volumes to obtain the volume of the metal.
3) For blocks:
   a) Measure the length, width, and height of the object as accurately as you can.
   b) Multiply these values together to find the volume.
4) Calculate density by dividing the mass by the volume. Keep 2 decimal places.
5) For the cylinder, pour out the water to reclaim the metal sample and dry it.
6) Repeat this process for the same metal two more times.
7) Once all trials have been completed for the one metal, obtain another sample and repeat the process.
8) Using the densities you calculated, determine the average density for each metal.
9) Use the average density and compare it to the chart above to identify the unknown metal.

Unknown Cylinder	Mass (g)	Volume			Density	Average Density	Name of Metal
		Initial Volume	Final Volume	Volume of Metal			
A							
B							
C							

Unknown Block	Mass (g)	Volume				Density	Average Density	Name of Metal
		Length (cm)	Width (cm)	Height (cm)	Volume of Metal			
D								
E								
F								

Analysis

1. Did the densities you calculated exactly match the ones in the chart on page 1? If not, why do you think that is?

2. What are some other observable properties, aside from density, that can be used to identify the metals?

3. Determine the percent error for each average calculation. Use the formula below. The measured value is the one you calculated. The accepted value is from the data table on the first page.

$$\% \text{ error} = \frac{|\text{measured value} - \text{accepted value}|}{\text{accepted value}} \times 100$$

A	
B	
C	
D	
E	
F	

4. Do you think density is a reliable method for identifying all unknown substances? Explain your answer.

5. Many substances expand when they are heated. How would this affect the density?

6. Copper has a density of 8.96 g/mL. What would the mass be if its volume was 34.7 mL?

7. You discover a piece of jewelry that you think might be pure silver. Using a balance, you determine it has a mass of 12.2 grams. With a graduated cylinder, you find its volume to be 1.9 mL. Is it likely silver?

Boat Lab – Teacher Notes

This lab begins with the students writing a research paper about boats and how they float. I generally give the students three class periods to write the 500+ word essay. This will give them some knowledge about first boats, famous boats, parts of a boat, etc. It also provides them with background information into buoyancy.

After the research is complete, I hold a class discussion about the lab we are about to perform. This can take half to a full period. If the discussion goes quickly with the students, then we start creating the boats that day. Otherwise, it waits until the next class. The information for the class discussion is included here, and I recommend converting it into a slideshow to focus the conversation.

Time: The opening discussion and building and measuring the boats takes one period. I use three periods to test each type of water. I then give one period for the students to work through the questions. We finish with a class discussion about the results.

Materials per group

- 1 square plastic tub (about 8" wide and long, able to hold about 3 cups of liquid
- Aluminum foil (4 sheets of 15 cm square; this is about half the width of a standard sheet)
- Pennies (100 per group in a plastic cup – easily obtained from a bank)
- 150 g salt (have extra on hand)
- 250 mL beaker
- Stirring stick
- Ruler
- Sharpie

Notes

Students are to make four differently shaped boats. They must record the surface area of the bottom of the boat to the best of their ability. Formulas are provided on the lab sheet. Once each boat is created, a student should write his or her name inside the boat with the sharpie. The boats will be needed for multiple days. If a boat crumbles (and it's likely) they can make another one of the same shape. Your pennies will end up salty.

To the class

Today we're starting the boat lab. For this, you will make four boats of different shapes and you'll place them in water to test their buoyancy. Remember, buoyancy is an upward force in a medium that keeps an object floating. We're measuring buoyancy with pennies. The more pennies a boat can hold before sinking, the higher its buoy145
ancy rating. When placing pennies into your boat, be careful. You don't want to accidentally sink the boat. If your boat buckles and touches the bottom, consider it sunk. We will test fresh water with no salt, an estimation of ocean water with some salt, and an estimation of saturated water with more salt. We want to compare (1) if boat shape has an effect on buoyancy and (2) if salinity has an effect on buoyancy. You need to save your boats at the end of your trials.

Discussion

Gather the student data and discuss the results. Did the shape of the boat matter? Did surface area work better? What happened with salinity? Note: results will be wildly skewed based on how the boats were handled. This is a good talking point. Also, discuss other possible errors (like dirty pennies, different amounts of water, etc.)

Name Period Date Teacher

Boat Research

For this assignment, you will do some research on the history of boats, parts of a boat, why boats float, types of boats, and other pertinent information.

You need to write a 500+ word essay explaining the details of boats. This is an independent research task. You can share some information with a friend, but your essays must be unique. Of course, you must write the essay in your own words and not copy/paste information from your sources.

Things to include in your essay:

- Several details regarding the history of boats
    - What are some of the first boats?
    - What are some boats of today?
    - What are some famous boats?
- Famous boating explorers
    - Include a minimum of three explorers
- Why do boats float?
    - Be sure to discuss buoyancy and forces
- What is the salinity of water in the ocean?
    - In the Great Salt Lake in Utah?
    - In the Dead Sea?
- Types of boats
    - Include at least three (i.e., trireme, sailboat)
- Parts of a boat
    - Include information where parts of a boat are specific to a type of boat (i.e. sails, motors)
- Three other facts or personal experiences you feel are important
- At least 3 internet sources
    - Full links, please
    - Copy and paste links from the address bar as you find sources

When you are finished, make sure you check over the requirements. Did you include your sources?

Boat Lab – Discussion

Problem

- Explore the effects of boat shape on buoyancy.
- Phrase this as a question
  - How does boat shape affect buoyancy?

Research

- What is a force?
  - A force is a push or a pull.
- What is buoyancy?
  - Buoyancy is the ability or tendency to float in water or air or some other fluid.
  - Think of buoyancy as an upward force that keeps an object floating.
  - We're going to use water.
- What type of boat shapes can we use?
- What are our boats made out of?
  - Aluminum foil
- How are we measuring buoyancy?
  - Number of pennies it can hold before sinking

Hypothesis

- What would be a good hypothesis for this experiment?
  - The ______ boat will have the most buoyancy.
    - square, triangular, round, etc.

Experiment: Independent Variable

- The shape of the boat you make
- Test each shape 3 times
- We're making boats out of aluminum foil
- Second independent variable
  - Testing salinities

Experiment: Dependent Variable

- The buoyancy
- We're measuring buoyancy based on the number of pennies the boat can hold before it sinks.
- This relates to the weight added. Weight is a force.

Experiment: Constants

- The amount of aluminum foil for each boat
  - Same weight for the boats
  - Same wall height between boats
- Water used per set of trials
  - same temperature
  - same amount of water
- Amount of force used to add pennies
- Penny placement
  - all in center? around edges? evenly spaced?

Experiment: Control Group

- Create a square boat.
- All other boats will be compared to the square boat.

Experiment: Experimental Groups

- Create at least three other boat shapes that aren't square.
- Remember your constants and use the same amount of aluminum foil and keep the wall height consistent.

Procedure

Setup

1. Get four sheets of aluminum foil
2. Cut (or rip) them down to the same size as needed
3. Decide on a shape for your boat
4. Bend the foil into the new shape
5. Remember to fold up walls on the edges
6. The boat must fit in the container without hitting the walls
7. Measure / calculate the surface area of the bottom of the boat
8. Record the surface area in your data table
9. Create three more boats

Action (After all boats are created and measured)

1. Fill container with 2 cups (475 mL) of water
2. Place one boat in the water
3. Add pennies until the boat sinks (touches the bottom)
4. Be sure to save your boat
5. Record the number of pennies
6. Repeat for at least three trials

Experiment: Data Table

Boat Shape	Surface Area	Fresh Water				Ocean Water				Saturated Water			
		Trial 1	Trial 2	Trial 3	Average	Trial 1	Trial 2	Trial 3	Average	Trial 1	Trial 2	Trial 3	Average

Analysis

- Which boat shape worked the best?
  - Why do you think so?
- Which shape performed the worst?
  - Why do you think so?
- How did each shape compare to the control group?
- Can you identify a trend?

Conclusion

- How does boat shape affect buoyancy?
- Give a brief summary of the experiment
- What are three possible errors?
- How could the experiment be improved?
- If you were going to build a boat, what shape would you use and why?

Names Period Date Teacher

Boat Lab

Despite their mass, boats are able to float on water, whether on a lake or an ocean. A rock with the same mass would sink, however. There are several differences between boats and rocks, including the shape of the object and the material itself. Buoyant forces act upon an object to attempt to keep it afloat, whether in the water or the air. In this lab, you will explore buoyant forces.

Problem: How do surface area and salinity affect buoyancy?

Research:
What is a force?

What is buoyancy?

Hypotheses: What impact will each have? ("The ___ will [increase/decrease/not affect] the buoyancy.")

Surface area:

Salinity:

Experiment: Part A: Surface Area

1) Gather four sheets of aluminum foil that are 15 cm x 15 cm in size. Please do not be wasteful.
2) Fold each sheet into a boat with a different surface area for the base. Different shapes are fine.
   a) Make sure your boats have walls.
3) Measure and calculate the surface area of each boat and record the data to the nearest cm^2.
   a) Try to record surface areas in the data table in order from lowest to highest.

Shape	Formula
Square/rectangle	length x width
Circle	πr^2

Shape	Formula
Triangle	½ base x height
Star	0.44 x width x height

4) Obtain a tub and fill it with 475 mL (2 cups) water. It does not have to be exact.
5) Obtain a container of pennies and some paper towels.
6) One at a time, place each boat into the tub and add pennies until the boat hits the bottom.
7) Record the number of pennies. This will represent buoyancy.
8) Repeat each trial at least twice to ensure the data is accurate. If results vary, test again.
9) When finished, ***save your boats***. You will need them again. Stack them on a sheet of paper with your name(s) and find a place to store them until the next class. Clean up your station.

Experiment: Part B: Ocean Salinity

Based on your research, what is the percent salinity of the oceans?

1) Obtain a 250 mL beaker from the drying rack in the back of the class.
2) Turn on a balance and make sure it is set to measure in grams (use the mode button if needed).
3) Tare the beaker by placing it on the balance and pressing the "on" button again to set it to zero.
4) Add 25 grams of salt to the beaker.
5) Fill your tub with 475 mL (2 cups) of water.
6) Add the salt to the water.
7) Stir until most of the salt dissolves.
8) Try to keep the salty water off the table.
9) Test your boats. Repeat steps 6-9 from Part A.

Experiment: Part C: Salt Saturation

According to basic solubility, it is possible to create a saturated solution by dissolving 35 grams of sodium chloride (NaCl) in 100 mL water. We're using a similar amount.

1) Add 100 grams of salt to the existing solution (or if you're making it fresh, use 125 grams)
2) Repeat the procedure for Part B.

Data Table:

Boat Shape	Surface Area	Fresh Water				Ocean Water				Saturated Water			
		Trial 1	Trial 2	Trial 3	Average	Trial 1	Trial 2	Trial 3	Average	Trial 1	Trial 2	Trial 3	Average

Analysis:

Questions: Part A

Which surface area had the greatest buoyancy? [smallest] [medium] [largest] [basically all the same]

Was your **surface area** hypothesis correct? Support with data.

Did the shape of the boat matter? Explain.

Questions: Part B

Which surface area had the greatest buoyancy? [smallest] [medium] [largest] [basically all the same]

Was your **salinity** hypothesis correct? Support with data.

Did the shape of the boat matter? Explain.

Questions: Part C

Which surface area had the greatest buoyancy? [smallest] [medium] [largest] [basically all the same]

Was your **salinity** hypothesis correct? Support with data.

Did the shape of the boat matter? Explain.

How did the increased salinity affect the buoyancy compared to Part B?

Graph

From the data, create a graph of the information. Place surface area along the bottom in order from lowest to highest. Place buoyancy (number of pennies) on the side. Plot the points for Fresh Water, then connect the dots from left to right. Add a second line for Ocean Salinity, and a third line for Saturated Salinity.

Conclusion:

1. Give a short summary of the process you followed during this lab. A few sentences will suffice.

2. Compare the surface area to buoyancy. Were there any patterns that occurred?

3. Compare the salinity to buoyancy. Were there any patterns that occurred?

4. Were these results what you expected? Why or why not?

5. Sometimes in a lab situation, unexpected results happen. What is a possible reason you or someone else may have seen unexpected results? What could have gone wrong?

6. What are some possible errors that occurred during the lab?

7. How could this experiment be improved or altered? What else could be tested?

CHAPTER 3

Molecular Properties

Molecules interact with each other in fascinating ways. They can cling together, repel each other, bond together, conduct electricity when dissolved, and so much more.

In this chapter, we start by exploring surface tension and the effect of a surfactant. We also explore solubility, saturation, and crystallization. We touch upon colligative properties and heat equations. Conductivity of ionic compounds is also on our list, as well as exploring water as the universal solvent.

Some of these labs take a fair amount of setup, so be sure to check the requirements first. Others will require a fair amount of material, too, particularly table salt.

Water on a Penny Lab – Teacher Notes

Students will use droppers to see how many drops of different liquids will fit on a penny before spilling off. This is often done in younger grades with just water, so this adds more materials to test, as well as a second independent variable: the use of a surfactant.

<u>Time</u>: One class period

<u>Materials per group</u>

- Test tube rack
- 4 test tubes, each with a liquid
  - Water, rubbing alcohol, vinegar, oil (or any liquids you want)
- 5 droppers (one for each test tube and one for soap)
- 1 clean penny
- Access to liquid soap (a squirt in a small cup is fine)
- Paper towels

<u>To the class</u>

Today we're going to test a property known as cohesion. Cohesion is the ability of molecules to cling together. This is different that adhesion, where molecules stick to other surfaces. You may have done something like this before, but we're jazzing it up. You're going to see how many drops of a liquid will fit on the surface of a penny. You'll be testing four different liquids today. In part B of the lab, you will also be adding a surfactant. A surfactant reduces the surface tension of a liquid. To do this, you will place a drop of soap on top of the dried penny before adding the droplets of one liquid. Then you'll repeat for the other liquids.

<u>Discussion</u>

- Go through the questions from the lab sheet
- Which liquid showed the highest cohesion (highest number of drops)? [usually water]
- Did you notice anything strange about the rubbing alcohol? [it might have evaporated before they could finish]
  - Alcohol has low cohesion, so it vaporizes quickly.
- What effect did the soap have? [should have reduced the number of drops]
- Which liquids are most likely to form spherical droplets? Why?
  - Liquids with higher cohesion will hold themselves together in a sphere

Names Period Date Teacher

Water on a Penny Lab

Part 1

Define cohesion:

Procedure A

At this station, you will place drops of liquids onto the surface of a penny.

1) Predict how many drops of each liquid the penny can hold.
2) Use a dropper to add one drop at a time on top of the penny. Count the drops.
3) When the liquid spills off the penny, stop and report the number of drops.
4) Run each trial twice and use the average for your comparisons. If your two trials are vastly different, run a third trial and discard the errant data.
5) Wipe off the penny after each trial.
6) Test each liquid. Do not mix the droppers between samples. Do not test the soap.

Liquid Without Soap	Number of Drops			
	Prediction	Trial 1	Trial 2	Trial Average

Part 2

Define surfactant:

Procedure B

1) Add one drop of soap to your finger.
2) Smear the soap over the top of the penny.
3) Repeat the process from above with each liquid. Add soap between trials.
4) Complete the data table on the back.

Liquid With Soap	Number of Drops			
	Prediction	Trial 1	Trial 2	Trial Average

Analysis

1. How did your predictions match with the results?

2. List the plain liquids without soap in order from least to most cohesion.

3. List the liquids with the surfactant in order from least to most cohesion.

4. How did the results compare with and without the surfactant?

5. Based on your data, what does a surfactant do to cohesion?

6. Which liquids are most likely to form spherical droplets? Why?

Conductivity of Ionic Compounds – Teacher Notes

Metals are good conductors of electricity, but so are ionic solutions to a different degree. We're testing the conductivity of various compounds to see which ones are ionic and which are covalent.

Time: One class period

Materials

- Samples to test per class
  - Each of these should be in a marked beaker or clear cup
    - I generally set up two sets to be shared by the class
  - Saltwater, tap water, deionized water, potassium chloride water, sugar water, baking soda water, vinegar, Essentia water, lemonade
    - Feel free to substitute any liquids
- Battery setup per group
  - Preferred: 9V battery with cap and wires; extension wires if needed
  - Alternate: 3 AA batteries, taped together end for end, aluminum foil to replace wires
    - If using this method, students will generally need to squeeze the batteries together to see the results; test saltwater first to see the bubbles

Notes

It is imperative that students do not allow the two electrodes to touch each other. This will short out the battery setup and could rupture the battery. This should be emphasized as part of lab safety.

The effect you're looking for is bubbles forming around one of the electrodes (wires) in the sample. I recommend having extra saltwater samples so students can test that first to see the control group.

To the class

Today we're going to test the conductivity of various compounds. Some compounds split into ions when they're dissolved in water. These ions can carry electricity through water. These are known as ***electrolytes***, You'll be creating a small electrical circuit and using it to test each sample. You'll attach the battery to the cap and then attach wires so they can reach into the sample container. Test out saltwater first so you get a sense of what you're looking for. You will be sharing solutions with your classmates, so only bring one sample to your station at a time, then return it so others can use it. At all times, make sure you keep the electrodes – the wires – from touching while they're hooked up to the battery. If they touch, they will burn out the battery and could harm you.

Discussion

Talk about the results with the class over their findings. If something bubbled unexpectedly, like sugar water, discuss why that might be. Maybe it was contaminated, for instance. Different liquids should have shown different amounts of bubbles, suggesting their strength of ionizing in water. Which samples bubbled the most? [Should be table salt and potassium chloride as both are salts that dissociate easily.] Which samples bubbled the least? [Deionized water and sugar water] Which samples were mild conductors? Were any a surprise?

Conductivity of Ionic Compounds – Teacher Notes

[illegible]

Name Period Date Teacher

Ionic Compounds Research

Use the internet to research the following questions.

Define the following terms:

Term	Definition
Crystals	
Cleavage planes	
Electrolyte	

1. Describe the melting and boiling points of ionic compounds.

2. Describe how lithium and chlorine form crystals.

3. What does it mean that ionic compounds are hard and brittle?

4. How is it possible to break apart an ionic compound?

5. What happens to ionic compounds when they are placed in water?

6. What happens to pure salts when they are melted?

7. How does a battery work, in short?

8. What is a problem with ordinary batteries?

9. What is special about rechargeable batteries?

10. What battery compounds produce lithium ions?

11. What happens when a lithium-ion battery charges up?

12. How do ions and electrons flow as compared to each other?

13. What happens to electrons if ions stop moving through the electrolyte?

14. What is a danger if lithium-ion batteries were to overheat and overcharge?

15. What is the "memory effect" of NiCd batteries?

16. Who invented lithium-ion batteries and when?

17. What are some advantages of lithium-ion batteries?

18. What are some disadvantages of lithium-ion batteries?

Names Period Date Teacher

<u>Ionic Compound Conductivity Lab</u>

Background: Ionic compounds are made when metals and nonmetals transfer electrons between them. This gives the metal a positive charge and the nonmetal a negative charge. When ionic compounds are in solution (i.e.: in water) they break into their separate ions. These allow for electrons to flow in the solution. When electrons flow, they create electricity. Therefore, ionic compounds are able to conduct electricity when they are in solution.

Task: You will test various solutions to see if they contain ionic compounds. By setting up a basic series circuit, you will test the liquids and look for the presence of a reaction, most notably in the form of bubbles.

Solutions to be Tested: Tap water, vinegar, saltwater, sugar water, water with sodium bicarbonate, potassium chloride, and possibly other solutions

Materials: In addition to the solutions listed above, you will need a 9V battery, a battery connector with wires, two additional wires (if needed)

Setup: Attach the 9V battery to the cap with the wires. Attach longer wires to the ends if needed. These electrodes will go into the solution you are testing.
DO NOT ALLOW THE ELECTRODES TO TOUCH EACH OTHER.
This will burn out the battery and make it extremely hot.

Procedure

- Obtain one sample solution. You do not need to do these in the order listed in the data table. It is recommended you start with the salt solution if possible. This is the control group.
- Predict whether you think the solution will conduct electricity. Record your guess in the data table.
- Set the two electrode ends into the solution, making sure the electrodes themselves do not touch.
- Observe for a minute or two to see if there is a reaction.
- Record your observations and determine if the solution is ionic.
- Return the sample to the station and obtain a new sample. Repeat the procedure until you have tested all samples.
- When all samples have been tested, disconnect the wires and the battery.
- Answer the analysis questions.

Solution	Prediction: Will it conduct?	Observations	Result: Based on obs., is it ionic?	Is it ionic? (Analysis Q #3)
Saltwater				
Potassium Chloride (KCl)				
Vinegar				
Tap Water				
Sugar Water				
Baking Soda Water				
Borax Water				
Lemonade				
Deionized Water				

Analysis

1. What makes a compound ionic?

2. How close were your predictions to the actual results?

3. Do a quick internet search to see if each of these compounds is ionic. Record the information in the data table above in the far-right column. Add a check mark if the data agrees with the research. (Ex: "Is sugar water ionic?" If it says "covalent" then it is not ionic.)

Colligative Properties – Teacher Notes

Colligative properties depend on the ratio of solute to solvent, not on the nature of the chemicals. We use salt to melt ice in the winter because of freezing point depression. Any material will lower the freezing point of water, but certain salts work better than other materials at reasonable cost and availability.

There are four main colligative properties.

- **Freezing point depression**: stronger solutions decrease the freezing point
- **Boiling point elevation**: stronger solutions increase the boiling point
- **Osmotic pressure**: stronger solutions decrease ability of a solution to permeate a membrane
    - Low concentration flows into high concentration
- **Vapor Pressure**: stronger solutions decrease the vapor pressure

Time: One class period

Materials per group

- Unfrozen freezer pop per person
- Two sealable plastic bags
- Ice
- Salt (50 g, 100 g, 150 g, 200 g, 250 g, depending on the group)
- Thermometer
- Small bathroom cup
- One large cup to get ice (or scooper in the freezer)

For salt, you'll need up to 250 grams per group. One group uses about 50 g, then 100 g, 150 g, 200 g, 250 g. That's about one kilogram of salt for one class. You can reduce these amounts by half if the salt supply is a problem. You may want to have students measure the masses on a balance rather than using a bathroom cup as a measuring device. You can also skip groups 4 and 5 and have other amounts tested twice.

You can obtain the freezer pops online if they're out of season when you're running this lab. I recommend splitting these apart before the lab, as the plastic tends to rip and can spill the juice. Students aren't as likely to tear the plastic patiently, thereby wasting a good number of ice pops. You may also want to have extra plastic bags on hand in case the two bags being used rip through (it happens). You will need a fair amount of salt for this lab, especially with the larger groups needing 5 or 6 scoops of salt. I don't always run a group zero, though it does still work; if I do, I usually do it myself. For ice, each group will need two large cupfuls or scoops. If you don't have an ice maker in your lab classroom, you can bring in a bag of ice or ask your athletic trainer if they have some.

To the class

Today we're going to test out freezing point depression. Each group will be placing unfrozen freezer pops – one per person – into a bag with a certain amount of salt. The freezing point of the ice will be lowered and the ice pops will freeze at different temperatures. You're going to be doing a lot of shaking in his lab, so get those hands ready. You're going to place the salt and pops into a plastic bag with some ice. Obtain the temperature as accurately as you can and record it as the temperature for time zero. Then seal the bag and stuff it into a second plastic bag. Now shake for two minutes, then check the temperature again. Repeat this process until the ice pops are completely frozen and you record similar temperatures for two to three trials. You may not need all 20 minutes. Record your final temperature on the board with the amount of salt you used. You'll collect this data from all groups. From there, remove your ice pops and wash them thoroughly. Then open them and enjoy! Seal up the plastic bags and discard them. There are lab questions to answer too.

Discussion

Table salt can potentially lower the freezing point of water to -21 °C, It's unlikely you'll see this value during the lab, though some students will have results as low at -16 °C or -18 °C. Once the data is on the board, point out any overall trend that you see. There should be a pattern of more salt leading to lower temperatures. After all, there's only so much a cupful of salt can do. Sometimes this doesn't happen with the data, so a discussion as to why might be worthwhile. Experimental error? Opening the bags to obtain temperature? Not shaking enough? Hands on the bag melting the ice?

Name Period Date Teacher

Colligative Properties Research

Use the internet to find the answers to the questions below. Restate the answers in your own words without copying and pasting.

1. What is a colligative property?

2. What are four main types of colligative property and what do they mean?

   a.

   b.

   c.

   d.

3. When does a substance cause greater freezing point depression, when there are ions or no ions? Why is this true?

4. What does non-volatile mean?

5. What is the freezing point of water, as a temperature in Celsius?

6. Why do we put salt on ice in the winter? Your answer should relate to freezing point depression.

7. Why do we use salt on ice in the winter and not sugar?

8. Water boils at 100°C. At what temperature would it boil if 50 grams of salt was added? (less than 100°C, equal to 100°C, or more than 100°C)

9. Sometimes the temperature is too cold for even salt (NaCl) to melt ice, for example if it's colder than -18°C (0°F). In this case, calcium chloride ($CaCl_2$) is used because it lowers the freezing point even more. Why do you think that is?

10. Plot the points on the graph, then answer the questions.

Amount of Salt (g)	Freezing Point (°C)
0	0
15	-2
30	-5
45	-9
60	-14
75	-18
90	-18
105	-18

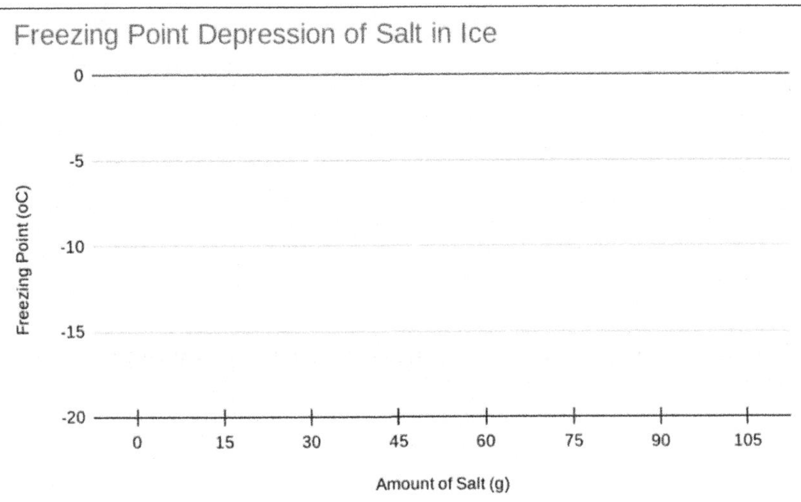

a) What is the overall trend of the graph?

b) What would you expect the freezing point to be with 52 grams of salt?

c) Once the amount of salt reached 75g, adding more didn't lower the freezing point anymore. Why do you think this is?

ESSAYS. Answer the following essays in your own words using at least 200 words for each essay.

1. Design an experiment to test freezing point depression using three different solutes. Identify the independent and dependent variables, control group and experimental groups, at least three constants, and at least three errors you want to control for.

2. How would you explain colligative properties to someone in 5th grade? You can focus on one property, such as freezing point depression. Use simple language so the 5th grader would understand it. Describe things the 5th grader might not know, such as what an ion is.

Names Period Date Teacher

Colligative Properties Lab

Purpose: To explore the colligative property of free point depression. When a nonvolatile solute (such as salt) is added to a substance, it causes the point of freezing to drop. Compounds that break into ions when dissolving are better at lowering the freezing point than substances that do not because the added particles disrupt the ability of water to form crystalline ice. The more particles there are, the greater the disruption and the greater the impact on particle-dependent properties such as freezing point. This lab will test this theory.

Materials: Unfrozen freezer pop, 2 sealable plastic bags, ice, salt, thermometer, small bathroom cup, one large cup (or scoop from the freezer)

Groups: All groups get 2 scoops of ice from the freezer
Group 0: no salt
Group 1: 1 cup of salt
Group 2: 2 cups of salt
Group 3: 3 cups of salt
Group 4: 4 cups of salt
Group 5: 5 cups of salt

**Note: a "cup" means a cupful, not a measured cup

Procedure:

1. Each group member obtains an ice pop
2. Gather the remaining materials. Place ice, salt, and ice pops in a sealable bag.
3. Stuff the filled bag into a second bag to prevent leaks.
4. Record the initial temperature at time zero.
5. Seal the bag.
6. Constantly shake the bag for 2 minutes.
7. Insert the thermometer into the bag and record the temperature and observations about the ice pop. Plot the temperature on the graph.
8. If your bag leaks, triple bag it with a third bag.
9. Repeat steps 4-6.
10. Record the final temperature on the board for comparison. Be sure to obtain the data from the rest of the class from the board.

Disposal:
Rinse the ice pop off thoroughly. Eat the ice pop! Throw out the ice pop wrapper. Drain all ice and saltwater into the sink. Rinse. Return the thermometer to the back lab station. Discard other materials.

If you do not wish to eat the ice pop, please discard it.

Data Table:

Amount of salt used (# scoops): ___________

Time (min)	Temperature (°C)	Graph
0		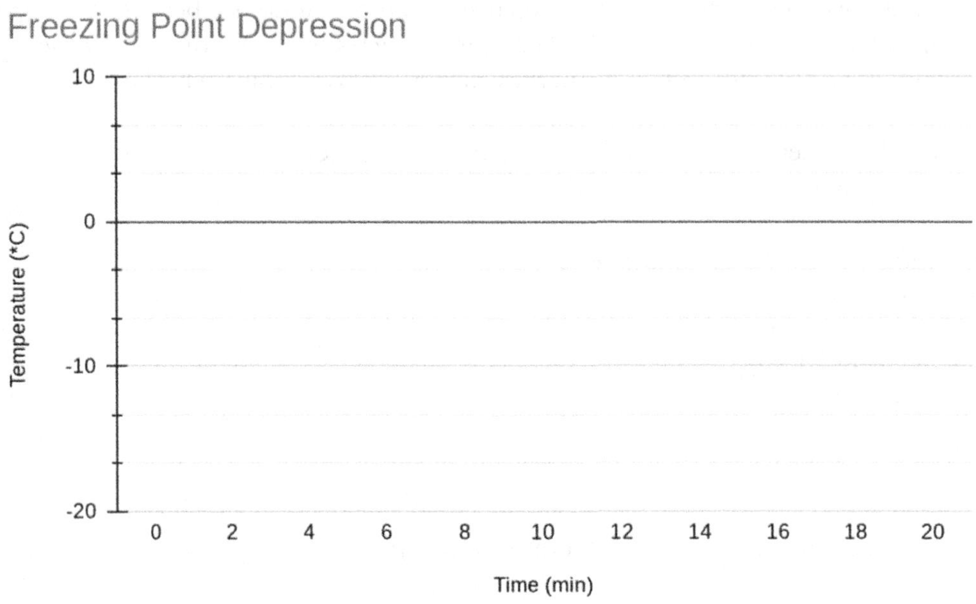
2		
4		
6		
8		
10		
12		
14		
16		
18		
20		

Group Data

Group #	Amount of Salt	Final Temperature(s)
0	0 scoops	
1	1 scoop	
2	2 scoops	
3	3 scoops	
4	4 scoops	
5	5 scoops	

Analysis:

1. What was the overall trend of your graph?

2. What was the overall trend of the amount of salt versus the final temperature (from group data)?

3. What would you expect to happen if 10 cups of salt were used for a trial?

4. Define freezing point depression.

5. Explain how this lab is intended to demonstrate freezing point depression.

6. Why do you think salt is put down on ice in the winter? How does it help?

7. Sugar does not break into ions, but salt does. What do you think its effect on freezing point depression would be as compared to salt?

8. What are three materials, aside from sodium chloride, that will lower the freezing point of water?

9. What are three things that might not be perfect about collecting data in this lab? (i.e. What are three possible experimental errors?)

10. Salt can potentially lower the freezing point of water to -21 °C. Assuming this as the accepted value, what is your percent error for this experiment? (This may be a large number.)

$$\% \text{ error} = \frac{|\text{measured value} - \text{accepted value}|}{\text{accepted value}} \times 100$$

11. Based on the group data, how could you alter your personal experiment to lower the percent error?

12. What would you change about this lab to make it better?

Heat Equations – Teacher Notes

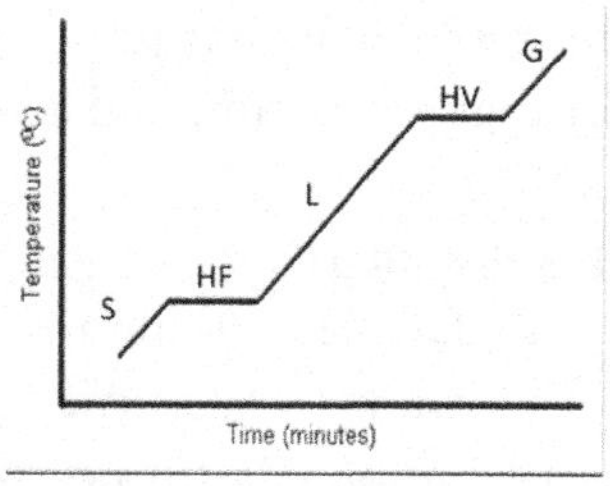

Energy is needed to increase the temperature of substances and to initiate certain phase changes. For water, it takes 4.18 J/g °C to raise the temperature of one gram of water by one °C. This is known as the ***specific heat capacity (c)*** and it can be used to calculate the energy needed to raise the temperature of liquid water. Each phase of water has its own specific heat. Ice, for instance is 2.095 J/g °C and steam is 2.0267 J/g °C. Note that these values are dependent upon pressure, so you may see slightly different values if you look them up elsewhere.

The formula needed to calculate the heat quotient (energy) needed to raise the temperature is as follows.

$$q = m\,c\,\Delta T$$

Where q is the heat quotient, m is mass in grams, c is heat capacity in J/g °C, and ΔT is change in temperature ($T_{final} - T_{initial}$). This formula is also used when cooling off a sample of H_2O; you will have a negative q.

You can also calculate the heat needed (or lost) during phase changes. To find the heat needed to melt ice, use the ***heat of fusion*** value of 334 J/g. To calculate the heat needed to boil water into steam use the ***heat of vaporization*** value 2260 J/g. These would also be used to calculate the heat lost when freezing water and condensing steam, respectively, but the q values will need to be negative.

$$q = m\,HF \qquad q = m\,HV$$

Note that you can also use kJ/kg °C or kJ/kg K as alternate units without changing the numerical values.

I start this lab with students performing some research about heat equations. They sometimes find some obscure equations, so a conversation later is important for understanding. They then conduct a lab to verify the heat of fusion of water as 334 J/g.

<u>**Time**</u>: One class period

<u>Materials per group</u>

- Two Styrofoam coffee cups
- Plastic cup
- Thermometer
- 100 mL graduated cylinder
- 250 mL graduated cylinder
- Water
- Ice cubes (handful)

Students will be placing 100 mL of water into the cup, then adding ice to melt. This will increase the volume of the water, obviously. That's the purpose of the 250 mL graduated cylinder; so they can obtain that new volume. If you don't have such large graduated cylinders, a simple adjustment is to pour the end product into the 100 mL graduated cylinder to the 100 mL line, empty it, then pour the rest in, adding 100 mL to the observed volume. This adds a little experimental error, but since the 250 mL graduated cylinder is less accurate anyway, it isn't a big deal.

This lab usually has good results. One issue is the removal of the ice cubes, which pulls water from the cups. But that's experimental error for you.

To the class

Today we're going to figure out how much heat is needed to melt ice. The accepted value is 334 J/g, and we want to see how accurate we can be to find the same thing. Today we're using Styrofoam cups, so please be careful not to crush or tear them. We'll be conserving them and I need to collect them back. Also, please don't leave the thermometer inside the cups, because it's most likely to knock over the cups.

Discussion

I suggest discussing the research task answers first before setting up and running this lab, as it will establish the background needed to perform the experiment with some knowledge. After the lab, discuss the results and the effectiveness of the process used to gather the data. Are there any better ways to control for errors?

Name Period Date Teacher

Heat Equations Research

Answer the following questions. Research answers on the internet. Include a list of sources at the end.

1. What is calorimetry?

2. What is a calorie?

3. Describe how to make a basic or normal calorimeter.

4. What is a bomb calorimeter?

5. Briefly describe the history of calorimetry. Who first invented it? When was it used? etc.

6. What is heat energy?

7. What is specific heat capacity? Also, what is the value of the specific heat capacity of water?

8. How, specifically, do you calculate change in temperature?

9. What equation is used to solve for the change of heat energy? What does each variable represent?

10. What is heat of fusion? Also, what is the value of the heat of fusion for water?

11. What equation is used to calculate the heat energy due to heat of fusion?

12. What is heat of vaporization? Also, what is the value of the heat of vaporization for water?

13. What equation is used to calculate the heat energy due to heat of vaporization?

14. How does heat of fusion change between the melting of ice and freezing of water?

15. What is a heating curve?

16. What does a general heating curve look like? Label its part, such as phases and heats.

17. At what temperature does water melt (in Celsius)? At what temperature does it freeze?

18. Based on what you've researched, how could you find the heat of fusion of an unknown substance?

Names Period Date Teacher

Heat of Fusion Lab

Purpose: The heat of fusion of a substance is the amount of heat required to fully turn 1 gram of a solid into 1 gram of liquid. For water, the value of the heat of fusion is 334 J/g. In this lab, we will attempt to calculate the heat of fusion for water using calorimetry.

Materials: Styrofoam coffee cups (x2), plastic cup, thermometer, 100 mL graduated cylinder, 250 mL graduated cylinder, water, ice cubes

Procedure:

1) Stack two coffee cups so one is inside the other
2) Use the third cup to obtain some ice cubes
3) Measure 100 mL of water (preferably room temperature) with the graduated cylinder
   a) If it's not at room temperature, let it sit for a few minutes first
   b) Record the exact amount of water as "Initial Volume"
4) Pour the water into the doubled-up coffee cups
5) Insert the thermometer and wait for the temperature to stop changing, then record it in the data table as "Initial Temperature"
   a) Be careful not to knock over the cups
6) Add 2-3 ice cubes to the cup, while stirring gently with the thermometer
   a) Do not let the ice cubes fully melt
   b) Continue adding more ice cubes, 2-3 at a time until the temperature stabilizes
7) When the temperature becomes constant, record the temperature as "Final Temperature"
8) Quickly remove the remaining ice cubes from the cup
9) Pour the water into the larger 250 mL graduated cylinder
   a) Take care not to spill!
   b) Record the volume of water
10) Return the cups and thermometer

Data Table:

Initial Volume of Water	
Final Volume	
Change in Volume (mass of ice)	

Initial Temperature	
Final Temperature	
Change in Temperature (ΔT)	

Calculations:

A. Calculate the change in temperature and record it here and in the data table. Delete any negative signs.

$\Delta T =$ ________________

B. Calculate the heat lost (q) using the formula $q = m\ c\ \Delta T$

- mass (m) = initial volume of water (because density of water is 1.00 g/mL, volume = mass)
- specific heat capacity (c) = 4.18 J/g °C
- change in temperature (ΔT) is from calculation A
- eliminate any negative signs

q = ________________

C. Find the mass of the ice by calculating the change in volume. Because water has a density of 1.00 g/mL, volume equals mass. Record this value here and in the data table.

$m_{ice} =$ ________________

D. Find the heat of fusion (H_f) of ice using the formula $q = m\ H_f$

- q = use the heat (q) found in calculation B
- m = use the mass found in calculation C
- use basic algebra to solve for H_f

$H_f =$ ________________

E. Calculate the percent error of the heat of fusion (H_f) using the accepted value of 334 J/g.

$$\% \text{ error} = \frac{|\text{measured value} - \text{accepted value}|}{\text{accepted value}} \times 100$$

% error = ________________

F. Give three possible reasons why the measured value and accepted value aren't exactly equal.

Solubility Lab – Teacher Notes

In this lab, solubilities for four solutes will be tested at different temperatures. Comparisons can also be made whether a solute dissociates or not, based on the data.

Time: three class periods plus one for discussion

Materials per group

- test tube
- flask or test tube rack
- thermometer
- chemical spatula
- test samples
- sheet of paper
- tray (optional)
- salt, sugar, sodium bicarbonate, potassium chloride samples

Notes

For setup, I fill plastic cups about halfway with each of the four solutes we're testing and cover them to help prevent them from absorbing moisture from the air. I do not make samples for every group, just maybe two sets for six groups. They will share the samples, but make sure you have extra solute on hand for when the sample cups get low. I recommend checking these before each lab period as needed.

The samples I use are salt, sugar, sodium bicarbonate, and potassium chloride. Feel free to substitute with other solutes instead, but keep the salt and sugar as they make great reference materials.

Keep extra test tubes on hand as well.

Students are going to make some mess in this lab as they try to pour solute from the spatula into the test tube. They should do this over the paper so the solute can be poured back into the sample cup before weighing.

There also tends to be some confusion about what to do with the thermometer. Students are to put the thermometer into the test tube while it's empty, then add the water up to the top. Removing the thermometer will leave an empty space to add solute into. Doing it this way, the water displacement of the thermometer will be a consistent way of making sure there's the same amount of water in the test tube for each trial.

You will need water samples at different temperatures. This is easy to achieve if you have a faucet with hot water. Students can simply mix cold and warm water to get different temperatures. If needed, use a hot plate to gently warm up some water. If so, students must wear goggles for safety. The water should only be heated until it is warm to the touch, not any hotter. If you have access to ice, this also helps provide a great range of temperatures. You won't need much ice each day. Just a beaker or two of water with a few cubes is sufficient. (Don't have an ice machine in your lab? Check with the athletic director. They often have ice on hand.)

Sugar should be able to dissolve about 10 grams, while salt will dissolve around 14 grams. Students will sometimes get values up around 20-25, which just means they spilled a bunch and didn't put it back in the cup. Those values should be discarded and retested if there's time.

To the class

Today we're exploring the effect of temperature on solubility, which is a measure of how much solute can dissolve in a solvent. We'll be testing four solutes at different temperatures. We will be running this lab over the next couple days.

[At this point, describe where the different temperature samples will come from, such as from the hot water faucet. Also, explain the use of the thermometer and water displacement.] Take care not to spill the solutes. If you do, pour it back into the cup ***before*** getting its mass. Usually, we don't put spilled chemicals back into their containers, so we're bending the rules on this one. If you get a really high mass, say around 20 grams, you should rerun that trial.

Discussion

I usually start with salt and create a data table on the board with students supplying their data through questioning. I'll ask for numbers in a range (say 0 to 10 °C) and write the values on the board in order of increasing temperature. We then evaluate the data to see if there are any erroneous measurements we should exclude. From here, I plot the points on a graph sketched on the board. I connect the points and show them that the average value is around 14 grams. I then repeat for sugar, which comes up around 10 grams. We discuss dissociation again, saying salt dissociates but sugar does not. Then we repeat the data collection for the unknowns and plot them. Based on where the data falls, we discuss whether those materials dissociate too. Usually, sodium bicarbonate comes in between salt and sugar, but closer to salt (it partially dissociates) and potassium chloride comes close to salt (so it dissociates). If you used a different solute, the data should lead you to the conclusion.

Based on the data, did temperature have a big effect on solubility for these solutes? Which material(s) dissociated? What are some other solutes can we try? (Consider running these if you want.)

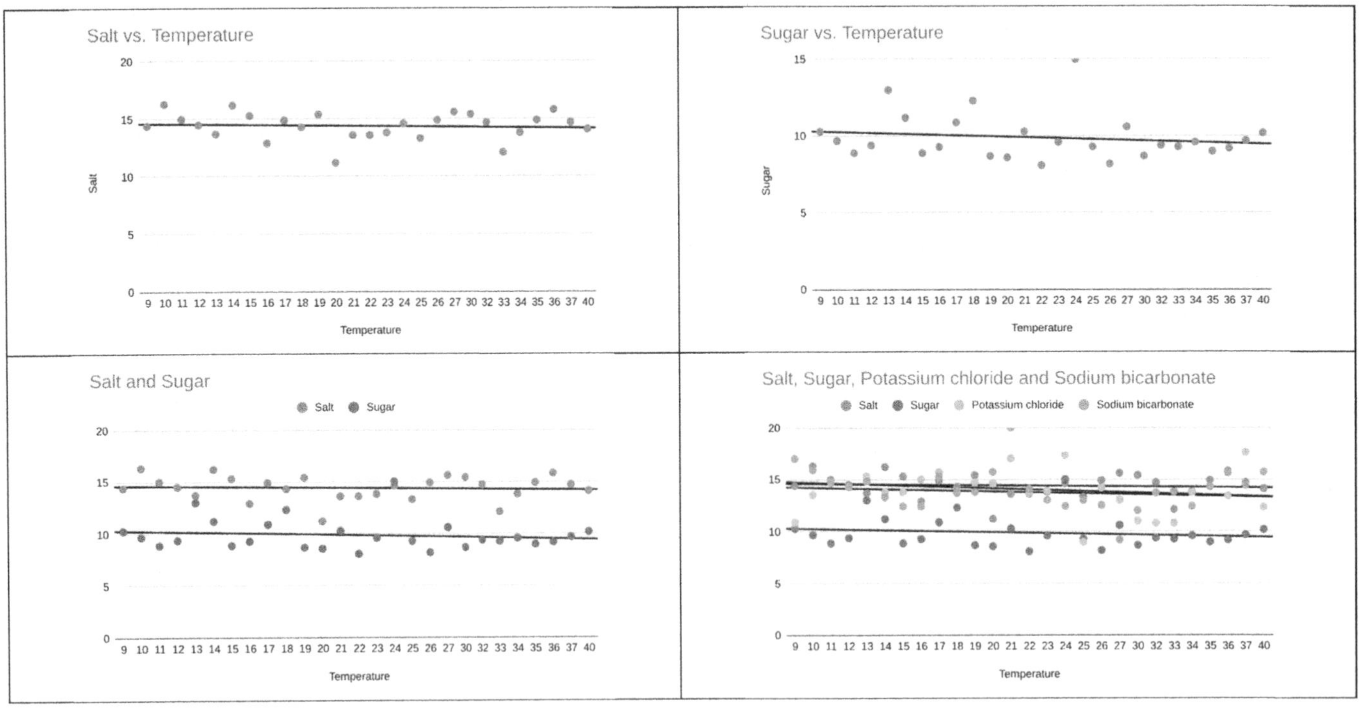

Averages: salt (14.4 grams), sugar (9.9 grams), sodium bicarbonate (14.1 grams), potassium chloride (13.8 grams)

Names Period Date Teacher

Solubility Lab

Purpose: Does temperature affect solubility? Does dissociation affect solubility? This lab answers these questions.

Materials: test tube, flask or test tube rack, thermometer, spatula, test samples, tray

Procedure:

1) Obtain one cup with a sample to be tested. Make sure the cup isn't nearly empty.
2) Find the mass of the cup with the sample and record it as the mass BEFORE.
3) Place the thermometer into the test tube.
4) With the thermometer inside the test tube, fill the test tube with a sample of water.
5) Record the temperature of the water.
6) Remove the thermometer.
7) Carefully set the test tube into the flask or test tube rack. Do not drop it in; the test tube will break.
8) Set the flask or test tube rack on top of a sheet of paper or tray.
9) Use the spatula to carefully scoop out the sample and place it into the test tube.
    a) Add the solute until the water level reaches the top.
    b) Keep the spatula dry.
    c) If you spill any, pour it back into the cup.
10) Find the mass of the cup with the sample and record it as the mass AFTER.
11) Subtract BEFORE and AFTER to find the mass of solute used.
12) Clean out the test tube.
13) Repeat this at several different temperatures.
14) Once you have filled in the data table for the sample, return the cup and obtain a new sample.

Data Tables: BE CAREFUL TO RECORD THE DATA IN THE CORRECT TABLE.

SALT

Value	Trial 1	Trial 2	Trial 3	Trial 4	Trial 5
Temperature (°C)					
Mass of Cup BEFORE (g)					
Mass of Cup AFTER (g)					
Mass of Solute (g) [subtract]					

SUGAR

Value	Trial 1	Trial 2	Trial 3	Trial 4	Trial 5
Temperature (°C)					
Mass of Cup BEFORE (g)					
Mass of Cup AFTER (g)					
Mass of Solute (g) [subtract]					

UNKNOWN A

Value	Trial 1	Trial 2	Trial 3	Trial 4	Trial 5
Temperature (°C)					
Mass of Cup BEFORE (g)					
Mass of Cup AFTER (g)					
Mass of Solute (g) [subtract]					

UNKNOWN B

Value	Trial 1	Trial 2	Trial 3	Trial 4	Trial 5
Temperature (°C)					
Mass of Cup BEFORE (g)					
Mass of Cup AFTER (g)					
Mass of Solute (g) [subtract]					

Solubility Lab Analysis

1) Create a graph of your data for salt. Use one color for this.
2) On the SAME set of axes, graph your data for sugar. Use a different color for this.
3) On the SAME set of axes, graph your data for unknown A. Use a different color for this.
4) On the SAME set of axes, graph your data for unknown B. Use a different color for this.
5) Make a legend that shows which color is for which solute.
6) Don't graph any data points that have significant error. For instance, if most of the values for a sample are close to 11 g, then don't graph a value of 25 g if you have one.

Solubility Graph

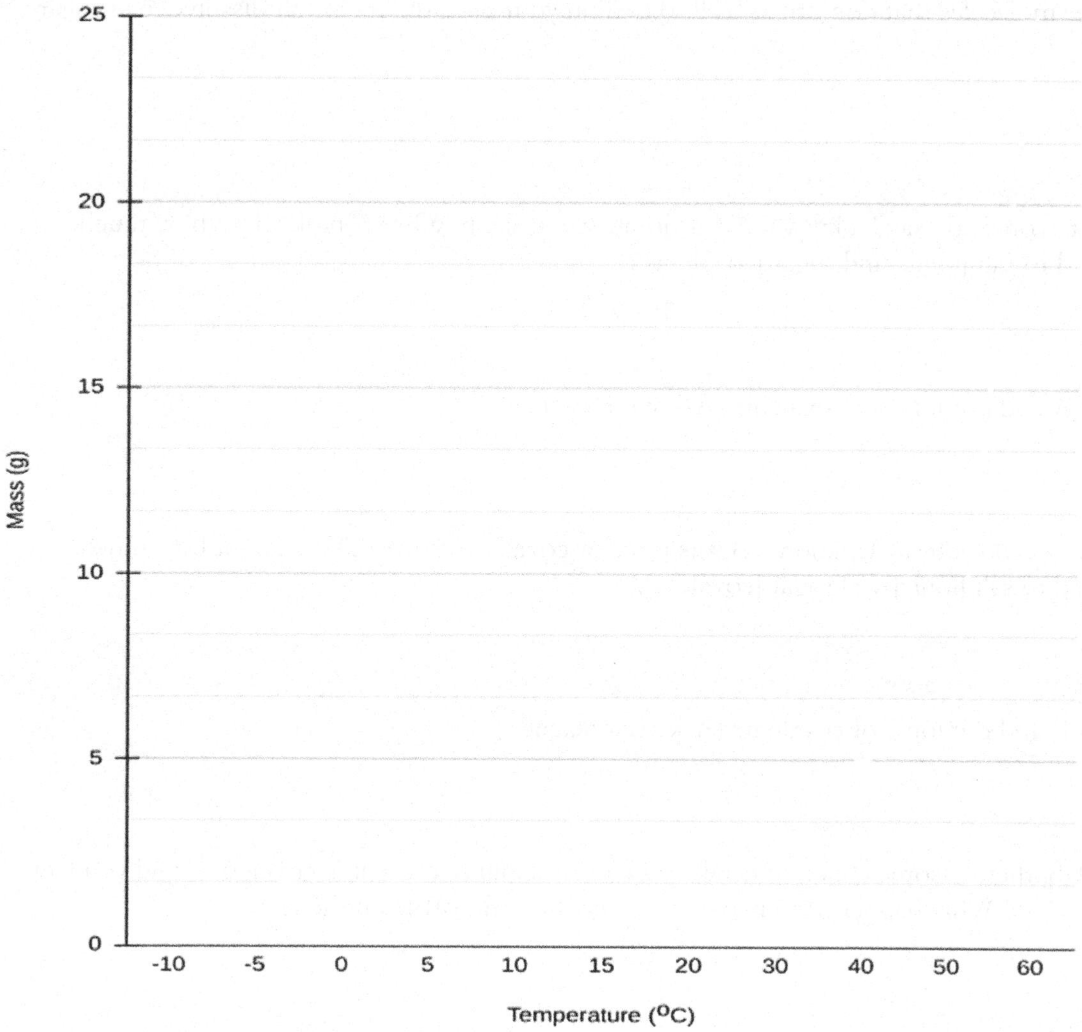

Legend: (circle with the appropriate color): SALT SUGAR UNKNOWN A UNKNOWN B

Averages. What was the average mass for each substance? Ignore any erroneous measurements, such as a 25 if all the other values are close to 11.

SALT	SUGAR	UNKNOWN A	UNKNOWN B

1. Salt is an ionic compound. It dissociates when it's in water. This means it breaks up into smaller ions. These smaller ions can fit better in between molecules. Sugar, however, is covalent and does not dissolve, and therefore has larger molecules that don't fit as well. Based on your data, did more salt dissolve than sugar overall?

2. Based on your data, would you say Unknown A was ionic or covalent? Why? (Think about how much dissolved compared to salt [ionic] and sugar [covalent].)

3. What is Unknown A and is it ionic or covalent? (Ask your teacher.)

4. Based on your data, would you say Unknown B was ionic or covalent? Why? (Think about how much dissolved compared to salt [ionic] and sugar [covalent].)

5. What is Unknown B and is it ionic or covalent? (Ask your teacher.)

6. Generally, in this lab, there is some data that needs to be thrown out because it is erroneous. Did you have any of this kind of data? What happened to make the numbers so far off from the rest?

7. What are three experimental errors that could have happened during this lab?

Concentration of a Salt Solution – Teacher Notes

Students take a salt solution of unknown concentration and try to determine its molarity. Some of the concepts in the analysis section of this lab may be a little advanced for a middle school group. This isn't the most intense lab. Much of the time is waiting for the water to boil and for the beaker and watch glass to cool.

Time: One class period

Lab safety
Students will be using hot plates, therefore goggles must be worn at all times. They should also be reminded not to touch the hot plate or glassware with their bare hands. They must use beaker tongs.

Materials per group
- Watch glass
- 10 mL graduated cylinder
- 100 mL beaker
- Hot plate
- Beaker tongs

Material for the class
- Salt solution prepared by the teacher
    - Each group needs 10 mL
    - Molarity should be different than 1 M

Background information
Gram formula mass (GFM) tells us the mass of a compound using the atomic masses from the periodic table. This gives us the number of grams of the substance per one mole. You must account for any multiples of atoms in a molecule, such as in potassium oxide, K_2O, which has two potassium atoms. On the periodic table, the mass of potassium is given as 39.0938. We'll round this to the nearest tenth: 39.1 g/mol. Because there are two potassium atoms, we multiply this by 2 to get 78.2. Oxygen's mass is 15.9994 (or 16.0), but because there's only one of them, we don't need to multiply it by anything. Once we have these two values, we add them together to get the gram formula mass: 78.2 + 16.0 = 94.2 g/mol.

Number of moles is a measurement of how much substance we have. It's like working with a dozen of something. If we have the mass of a substance in grams and we have the substance's gram formula mass, we can calculate the number of moles. All we do is take the actual mass and divide it by the gram formula mass. For instance, if we had 153.5 grams of potassium oxide, we would divide that by 94.2 to get: $153.5 \div 94.2 =$ 1.63 moles.

Molarity is the measure of a solution's concentration, or strength. This is represented by the number of moles per liter of a substance. The stronger the solution, the higher this number will be, because there will be more moles within a given volume. We calculate molarity by dividing the number of moles by the volume in liters. Let's say we have 8.6 L of potassium oxide. We take the moles (1.63 moles from above) and divide it by 8.6 L. This gives us: $1.63 \div 8.6 = 0.19$ M concentration.

To the class

We're using hot plates today, so lab safety is extremely important. Remember, the surface of the hot plate and the glassware on it will look the same hot or cold. You must never touch either of them with your hands. Goggles need to be worn throughout the lab.

I'm giving you a solution with an unknown concentration of salt. Your task is figure out what that concentration is. You'll need to measure out the solution accurately and record the precise volume. Is it exactly 10.0 or is it 9.9 mL? You need to be precise.

At one point, you will place your beaker on the hot plate with the watch glass on top. The watch glass must be set so it's like a bowl on top of the beaker, not a dome. This will keep it from falling off. When all the water had boiled off, you will need to use the tongs to remove the watch glass, then beaker, from the hot plate. Let them cool for about 10-15 minutes before continuing with the lab. Even then, you need to use the tongs.

Discussion

After discussing the background information with the students, you may also need to explain how to convert from mL to L. The analysis questions lead the students through the process of finding the unknown concentration one step at a time.

Possible sources of error could be the volume reading wasn't accurate, not all of the solution was poured into the beaker, some salt escaped the beaker, etc.

Names Period Date Teacher

Concentration of a Salt Solution Lab

Purpose: To determine the concentration of a sodium chloride solution using reduction to obtain NaCl solute. Concentration is expressed as moles per liter (Molarity) using the following equation.

$$Molarity = \frac{Moles\ of\ solute}{Volume\ of\ solution}$$

Equipment: watch glass, 10 mL graduated cylinder, 100 mL beaker, hot plate, beaker tongs

Raw Material:
NaCl of unknown concentration

Lab Safety:
Goggles must be worn at all times. Be conscious of hot metal and glass.

Procedure:

1) Obtain a clean, dry 100 mL beaker and a watch glass.
2) Record the mass of the beaker and watch glass together to the nearest 0.01 g.
3) Measure out 10 mL of the unknown NaCl(aq) solution using a 10 mL graduated cylinder.
    a) Record the actual volume in the data table to the nearest 0.1 mL.
4) Pour the NaCl(aq) solution into the beaker and set on the hot plate.
5) Cover with the watch glass with the convex side down, making a bowl shape.
6) Heat the solution until the water boils completely away. Continue heating for several more minutes.
7) Let the beaker cool for about 10-15 minutes.
8) Use the tongs to remove the watch glass, then carry the beaker and then the watch glass to the balance to obtain the mass to the nearest 0.01 g.
9) When the beaker and watch glass are sufficiently cool, rinse with water to clean.

Mass of Empty Beaker and Watch Glass	Mass of Beaker, Watch Glass, and Salt AFTER boiling	Mass of Salt (Subtract the other two masses)	Exact Volume of NaCl solution (mL)

Calculations:

1. Convert the volume of NaCl from mL to L. (Divide by 1000.)

2. Determine the mass of the salt (NaCl). (This is also in the data table.)

3. Calculate the gram formula mass (GFM) of NaCl to the nearest tenth.

4. Calculate the number of moles of NaCl in the solution. (mass of NaCl ÷ GFM)

5. Calculate the molarity of the NaCl solution using the formula on the reverse side. Be sure to use liters.

6. Ask your teacher for the actual (accepted) molarity of the solution and record it here.

7. Calculate your percent error using the formula shown.

$$\% \text{ error} = \frac{|\text{measured value} - \text{accepted value}|}{\text{accepted value}} \times 100$$

8. How can you account for the percent error? Give three possible reasons why the percent error is not 0%

a.

b.

c.

Crystallization – Teacher Notes

In this lab, make crystals using various solutes. This lab makes use of hot plates, so it's important to emphasize lab safety and for everyone to wear goggles. Students will especially want to take their goggles off while waiting for the water to boil. For safety, they must keep them on. Students will be stirring solutes into hot water, so it is vitally important they never touch the hot beaker with their hands. Emphasize this many times.

Time: six class periods, non-consecutive

Notes

I recommend running this lab once a week, each time with a new solute. This will give ample time for crystals to grow. I've used it on an every-other-day basis to some effect, too, but more time helps. If you find crystals aren't growing, try adding more solute. You can also add solute to the hot beakers until a small amount of precipitate forms. I haven't found this too necessary. I use 600 mL beakers, but 400 mL work too; I just find more crystals form on the sides than desired with the smaller beaker.

Depending on the age and maturity of your group, you may want to be responsible for all beaker movement.

I usually set up beakers with excess solute. It's important to remind the students not to use it all. They have to use the measured amounts given in the chart, which means they will have some extra solute when they're done.

It generally only takes about 20 minutes to set up each sample, which includes boiling the water. It takes about another 5 minutes or so to clean up the sample from the previous lab day. This provides extra time for students to potentially do research on crystal formation or watch videos curated by you about crystals. All in all, this lab with five solutes takes six periods to run. The sixth day is cleaning up the last solute and answering the analysis questions. You can, of course, test additional or alternate solutes. If so, you should research the saturation point of the solute in 100°C water so you know how much solute you need.

Clean up of most of the solutes is relatively easy. Some crystals, however, will stick to the sides of the glass and can be hard to remove. This is easily overcome by filling the beaker with water and boiling it, which you should do yourself. You may want to have extra beakers on hand for these instances so the students can still set up the next solute.

Materials per group per period

- Solutes: only one solute needed per group per class period
    - table salt (105g), sodium bicarbonate (35g), sugar (400g), Epsom salt (200g), borax (85g)
- Hot plate (therefore safety goggles)
- 2 wood sticks
- 1 half-length of a pipe cleaner
- 600 mL beaker (400 mL is okay too)
- Beaker tongs
- Magnifying glass (optional)
- 250 mL beaker or paper plate for measuring solute
- Chemical spatula

To the class

Today will be making use of hot plates, so goggles must be worn at all times, even when you're waiting for your water to boil. You must also NEVER touch a beaker with your hands once it has been on the hot plate. Use beaker tongs to move your beakers. I will move your beakers to and from the hot plates, but you'll be adding solute to the hot water. You must do this carefully. Take care not to splash, and remember the glass will be hot and can burn you. The sample beakers you're getting have more solute than you need, so make sure you measure carefully. You should end up with extra solute after you've measured accurately. Don't simply use the whole sample.

[When you're testing sugar…] With the sugar we're using today, measure out only 100 grams at a time. You should be able to dissolve about 400 grams in your water, but only add it little by little.

Discussion

- When a solution is "unsaturated," it is "not full," so it can dissolve more solute.
    - I think of it like packing for a vacation, but you only put two T-shirts and socks in the suitcase; there's room for more clothes.
- A "saturated" solution is "full," so it can't dissolve any more.
    - This is like filling the suitcase to the top before zipping it shut.
- A solution can also be "supersaturated" which means "more than full," so it has more solute dissolved than normal.
    - This is like taking that full suitcase and shoving an extra jacket inside, then you and a friend have to jump on top of the suitcase in order to zip it up. If you're not careful, you might cause that zipper to pop out and spill your clothes out. That's what we're doing in this lab.
    - We're dissolving more solute than normal, then it "falls out of solution" as it cools, forming crystals.

Additional Questions

1. Some of the crystals clung to the glass instead of the pipe cleaner. Why might this be? [beaker was a little dirty, so the crystals clung to those particles; adhesion with the glass: the water solution clung like a meniscus]
2. We only measured the mass of the crystals that formed on the pipe cleaner. How could we have measured all the crystals that were formed instead? [get the mass of the empty beaker, wait for crystals to form, pour out the liquid, then measure the mass again]
3. What was the point of dipping the pipe cleaner in the solute after wetting it? [to give seed crystals for the rest to cling to]
4. What are some other solutes we could try? [consider actually trying them]

Research: We start off with a research task regarding crystals.

Time: 2-3 days

Name Period Date Teacher

Crystals Research

What are crystals?

How do crystals form?

What unique properties do crystals have?

Are snowflakes crystals? How so?

What are timing crystals?

What are three facts about diamonds?

What are three details about quartz?

What is a geode?

What is the chemical composition and color of each of the following crystals? You're looking for the elements that make up the crystal or the chemical formula (like H_2O).

Crystal	Chemical Composition	Color
amethyst		
rose quartz		
malachite		
turquoise		
citrine		
onyx		
fluorite		
agate		
jasper		
opal		
topaz		
emerald		

Some people claim crystals can be used for healing. What are ten types of healing crystals and how are they supposed to help with healing?

Do you believe crystals can be used for healing? Why or why not?

What kinds of crystals form in caves? (i.e.: What materials are they made of?)

How do cave crystals form?

What is the difference between a stalactite and a stalagmite?

Where are some crystal caves around the world? List at least three.

Define the following chemistry terms.

Term	**Definition**
unsaturated	
saturated	
supersaturated	

How is a supersaturated solution created?

How is rock candy made?

Names Period Date Teacher

Crystallization Lab

Materials: 600 mL beaker, hot plate, beaker tongs, pipe cleaners, wood sticks, water, solutes, magnifying glass, 250 mL beaker or paper plate, chemical spatula

Lab Safety: Goggles must be worn. ABSOLUTELY DO NOT TOUCH THE HOT BEAKER WITH YOUR BARE HANDS!!! Once the beaker has touched the hot plate, assume it is hot. Use the beaker tongs to move the beaker when needed.

Procedure:

1) Take a pipe cleaner and attach it to the center of a wood stick: measure the mass
2) Place the wood stick across the beaker and make sure the pipe cleaner doesn't touch the beaker at all; if it does, reshape it
3) Wash the 600 mL beaker then add 300 mL of water
4) Prepare the solute:
   a) Place the weighing tray/plate on the balance
   b) Zero the balance (tare the plate)
   c) Weigh out the needed amount of solute (see chart) DO NOT ADD THIS TO THE BEAKER
5) Dip the pipe cleaner into the water, then into the solute, then set it aside to dry
6) Set the beaker on the hot plate and turn the hot plate on; have a seat until the water boils
   a) Keep your goggles on
7) When the water boils:
   a) Remove the beaker from the hot plate USING THE TONGS
   b) SLOWLY add the compound to the boiling water and stir carefully
   c) DO NOT TOUCH THE BEAKER WITH YOUR HANDS
8) Place the wood stick with the pipe cleaner across the top of the beaker
9) Set the beaker aside so it can cool down. DO NOT TOUCH THE BEAKER WITH YOUR HANDS.
10) After a few days, crystals should have formed. At this time, take the pipe cleaner out of the beaker
11) Determine the mass of the crystals plus pipe cleaner, then subtract the pipe cleaner's mass
12) Clean out the beaker, then repeat the experiment if there is a new solute

Data

Compound	Mass of Solute to Use	Mass of Pipe Cleaner	Mass of Crystals with Pipe Cleaner	Mass of Crystals (subtract)	Description of Crystals (How do they look?)
Table Salt	105 g				
Sodium Bicarbonate	35 g				
Sugar	Up to 350 g (Dissolve in 100g amounts)				
Epsom Salt	Up to 200 g				
Borax	85 g				

Analysis:

1. Which solute produced the most crystals?

2. Which solute produced the fewest crystals?

3. Did any crystals form in the beakers that did not form on the pipe cleaner?

4. How could the excess crystals in the beaker have been measured?

5. What's another solute that could have been tested? (You can use this internet for help.)

Universal Solvent Lab – Teacher Notes

Test different solutes in different solvents to see what combination dissolves the best.

To be honest, there is a fair amount of setup for this lab with a lot of graduated cylinders and beakers, all of which need to be labeled per group, which you'll see in the materials list. The reason for all these materials is so we don't contaminate each solvent being tested.

You can get by with only one of each beaker and graduated cylinder if you decide to keep a large supply at a "filling station" where students measure out the needed materials instead. I've done this to reduce setup. I found it important to separate each solvent by a good distance, as not everyone is known for organization and neatness.

Time: One class period

Materials per group

- All labeled A through D
  - 250 mL beaker for each solvent being tested, filled with 200 mL of each solvent
  - Four 100 mL (or larger) beakers
  - Four 50 mL or 100 mL graduated cylinders
- Solvents (200 mL each)
  - Water, rubbing alcohol, witch hazel, baby or vegetable oil
    - Feel free to substitute any liquids, but make sure you include water
- Solutes (4 teaspoons each – about 20-25 grams total per solute)
  - Salt, sugar, baking soda, powdered milk (this can be hard to see whether the solute dissolved)
    - Feel free to substitute any solutes
  - Four 50 mL or 100 mL beakers with each solute being tested, labeled with solute names or letters (not A through D)
- Stirring sticks or glass rods
  - I recommend 4 (one per solvent) – wipe off between trials
- Four spoons (one for each solute)

One note for cleaning up. You should have a waste container (2L bottle, for instance) for students to empty the oil into, so it can be sealed and discarded in the garbage, rather than be poured down the sink.

To the class

Water is called the "universal solvent" because it can dissolve a great many things, known as solutes. Today we're going to compare water's ability to dissolve certain solutes against other solvents to see which one dissolves the best. It's very important today that you don't mix liquids together. Everything is labeled A through D. Use graduated cylinder A with solvent A and pour it into beaker A, and so on. If you're not careful, you'll mess up the results. During the lab, you will be asked to empty the mixture beaker. When you're emptying out the oil mixture, pour it into the waste beaker, not down the drain. We're using a teaspoon today instead of measuring out the solute masses exactly. Use a level teaspoon each time so the masses will be approximately the same. I have included a set of placemats to help you stay organized.

Discussion

Water is the "universal solvent," and you should have seen that it dissolved more solutes than the other liquids. This lab is a little tricky in data collection, because it can be hard to see if all the crystals dissolved. So if you found a different liquid to be better, that could be why.

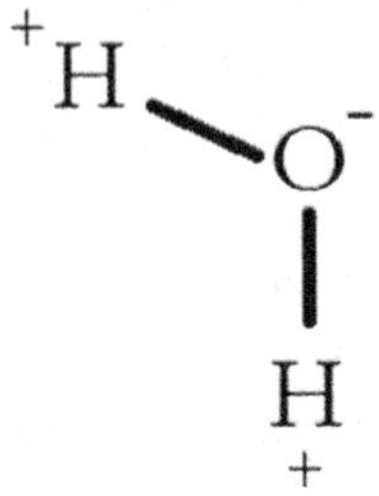

Why does water dissolve things so easily? It comes down to the shape of water molecules and something known as polarity. There is an angle of 104.5° between the hydrogen bonds. Oxygen has a higher electronegativity (affinity for drawing electrons closer: 3.44) than hydrogen (2.2). This means the electron from each hydrogen sits closer to the oxygen, giving oxygen a slightly negative charge. Because the electrons have shifted away from the hydrogens, this gives the hydrogen atoms a slightly positive charge. These differences in charges are known as ***polarity***. Water is a polar molecule. When two water molecules come close to each other, the slightly positive hydrogens of one molecule are drawn toward the slightly negative charge of the oxygen in the other molecule, forming a strong intermolecular bond known as a ***hydrogen bond***. Working together, these properties allow water to interact with many substances, drawing them in and dissolving them, making water the universal solvent.

Oil, on the other hand, is a larger nonpolar molecule. It's harder for it to interact with other materials, and you should have seen this as one of the lowest dissolving solvents we tested. Sugar is a nonpolar solute and should have dissolved more readily in the oil than in the other materials. The other solutes were polar and should not have dissolved well in the oil.

Rubbing alcohol and witch hazel have polar and nonpolar ends. They are able to form hydrogen bonds with water, but because of their molecule size, they "get in the way" of other solutes being dissolved, and therefore do not dissolve solutes as well as water.

Solutes (Powders)

	(E) Salt
	(F) Sugar
	(G) Baking Soda
	(H) Powdered Milk

Solvents (Liquids)

	(A) Water
	(B) Rubbing Alcohol
	(C) Witch Hazel
	(D) Baby Oil

Mixtures (Solute + Solvent)

	(A) Water + Solute
	(B) Rubbing Alcohol + Solute
	(C) Witch Hazel + Solute
	(D) Baby Oil + Solute

Solvents (Graduated Cylinders)

	(A) Water
	(B) Rubbing Alcohol
	(C) Witch Hazel
	(D) Baby Oil

Names Period Date Teacher

Universal Solvent Lab

Purpose: To test the solubilities of different solutes in various solvents.

Materials: Four beakers with solute (labeled A through D), four graduated cylinders (labeled A through D), four empty beakers (labeled A through D), stirring sticks, plastic teaspoon

Station Solubility

Part 1: Define solute:____

Define solvent:____

At this station you will test the solubilities of different liquids.

1) Predict which solvent(s) will dissolve the most. Record in the chart below.
2) Measure out 50 mL of each liquid into a separate cup or beaker.
    a) Use the labeled graduated cylinders with the labeled beakers. Do not mix them up.
3) Place 5 grams (one spoonful) of salt into each beaker.
4) Stir for a few seconds using a stirring stick (not the spoon).
5) Observe each mixture for 1 minute, then record how much dissolved.
6) Empty out the beakers.
    a) Oil gets emptied into the waste container
7) Repeat the experiment with other solutes (sugar, etc.)
8) Answer the analysis questions on the back.

Prediction. Rank the solvents in order from least to greatest. Which do you think will dissolve the most solutes?

Least	Middle		Greatest

Solubility.

Use this key: 3: nearly all dissolved 2: partially dissolved 1: not dissolved

Liquid being tested	Solute				Total Score
	Salt	Sugar			
Water					
Total Score					

Analysis:

1. Rank the solvents in order from least dissolved to most dissolved. (Hint: add up the values in each row.)

2. How did this match your prediction?

3. Water is considered to be the "universal solvent." It should be able to dissolve all the materials at this station. How close did your results match this description?

4. Which solute dissolved in the most solvents?

5. How could this experiment be improved? What other solutes and solvents could be tested? What other variables could be tested?

Average Atomic Mass (Beanium) – Teacher Notes

Atoms exist with certain isotopes. Isotopes are two versions of an element that have a different number of neutrons. Each isotope has a certain abundance in our world. For example, chlorine-35 has an abundance of about 75% and chlorine-37 has an abundance of about 25%. These are not exact values, but they demonstrate the concept. Each element has an average atomic mass, which is located on the periodic table. This average is determined by the abundance of all the element's isotopes. This is called average atomic mass.

To simulate the finding of an average atomic mass, students will determine the value of a fictional element called beanium, made up of three isotopes, redium, whiteium, and blackium. Once you set up this lab, you can use it over and over with no additional setup. Before you run the lab, do a test yourself and find the average mass. It's needed so students can calculate the percent error.

Time: One class period

Materials per group

- A baggie with:
  - four weighing trays (4 is a recommendation; 1 works too)
  - 27 red beans, 13 black beans, 10 white beans
    - You can use any amounts you want, just vary them
    - these should be dry beans
- balance

To the class

Atomic isotopes exist in certain abundances in the world. These abundances make up an average atomic mass. Carbon, for instance, has a mass of 12.011 u. 98.93% of all carbon is carbon-12, which is carbon with 6 neutrons. The rest is made up of carbon-14, which is carbon with 8 neutrons. These are isotopes of each other. Today, we're going to find the average atomic mass of beanium, a fictional element. For this, you will count out three types of beans and find their masses. You'll figure out their relative abundance, and use that to calculate the average atomic mass of beanium. Please take care not to drop any of the beans on the floor.

Discussion

We see in this lab that redium is the most abundant. There are more red beans than the others. This means it will contribute the most to the average atomic mass. The white beans had the fewest number, so their contribution is the least.

Why did you have some percent error? What caused your results to be different than the accepted value? [accuracy of the balance was too low for the samples, etc.]

Why weren't these considered perfect isotopes? [the beans were all different sizes, unlike neutrons, which are the same size]

Why would testing 1,000 beans be better than testing 50? [larger samples reduce error because of the accuracy of the balance; it also averages out the differences in the various bean sizes]

Names Period Date Teacher

Average Atomic Mass (Beanium) Lab

Introduction

Atomic mass of an element is the weighted average of all that element's naturally occurring isotopes. Isotopes of an element differ from one another based on the number of neutrons in the nucleus. The only difference between isotopes is the mass due to these neutrons. All isotopes of an element have the same number of protons, so the only thing that can change is the neutron number. A weighted average means that the relative abundance of each isotope must be factored into the calculation of the average.

Chlorine exists in nature as two isotopes, chlorine-35 and chlorine-37. Chlorine-35 has a mass of 35 u and a relative abundance of 75% and chlorine-37 represents the other 25% with a mass of 37 u. What is the average atomic mass of chlorine?

Step	What To Do	Chlorine-35	Chlorine-37
1	Write the mass of the isotope		
2	Write the percent abundance of the isotope as a decimal		
3	Multiply the mass by the percent		
4	Add the values from step 3 to get the average atomic mass of chlorine		

Percent Error

Calculate the percent error of the calculated (measured) average atomic mass of chlorine from step 4 versus the accepted value given on the periodic table (35.453).

$$\% \text{ error} = \frac{|\text{measured value - accepted value}|}{\text{accepted value}} \times 100$$

You'll notice there is some percent error from the calculation. This is because the percent abundance of chlorine-35 isn't exactly 75%. It's closer to 75.784%, but using 75% gives a close approximation.

Purpose of the Lab

You're going to determine the average atomic mass of an imaginary element called beanium. Beanium has three naturally occurring isotopes: blackium, redium, and whiteium. As in real life, the average atomic mass is calculated from the actual masses of the isotopes. Your job is to obtain a sample of beanium and determine the relative abundance of each isotope. Unlike real isotopes, there will be some variation in the masses of the sample isotopes. In other words, one red bean might have a slightly different mass than another red bean. This wouldn't be true in real isotopes. But that's experiments for you. From the data, you will calculate the average atomic mass of beanium.

Materials

Beanium sample, digital balance, weighing trays

Lab Safety

Handle your beanium isotopes with care. Do not drop them, lose them, or eat them.

Procedure

1. Count the number of each isotope in your sample. Make sure it matches the value on the board.

	Blackium	Redium	Whiteium
Number of each isotope (# beans)			
Total number of beans			

2. Determine the relative abundance of each isotope. To do this, divide the number of beans in each isotope by the total number of beans. Leave the answer in decimal form. (Don't multiply by 100.) Keep three decimal places.

	Blackium	Redium	Whiteium
Relative abundance			

3. Find the average mass of each isotope. To do this, get the total mass of one isotope and divide it by the number of its beans (from step 1). To get the mass, place the weighing tray on the balance and press the Zero or Tare button, then add the isotope's beans. Keep three decimal places.

	Blackium	Redium	Whiteium
Total mass of the isotope (g)			
Average mass of the isotope (divide the mass by the number of its beans)			

4. Calculate the average atomic mass of beanium. Do this by taking the relative abundance from step 2 and multiplying it by the average mass of the isotope from step 3. Finally, add those values together. In this chart, the first two lines ask you to **copy your data** from steps 2 and 3.

	Blackium	Redium	Whiteium
Relative Abundance (copy from Step 2)			
Average mass of isotope (copy from Step 3)			
Multiply the above values to get the mass contribution of each isotope			
Add the masses to obtain the average atomic mass of beanium. Keep three decimals. Do not divide by 3.			

Group Data. Ask around for the final answer from your other classmates. What average atomic mass did they find?

Group 1	Group 2	Group 3	Group 4	Group 5	Group 6	Group 7	Group 8

Percent Error

Calculate the percent error of your measured average atomic mass of beanium from step 4 versus the accepted value, which is 0.2022.

$$\% \text{ error} = \frac{|\text{measured value - accepted value}|}{\text{accepted value}} \times 100$$

Analysis

1. Which pieces of your data can be measured and which can be calculated?

2. Explain differences between the atomic mass of your beanium sample and that of your classmates.

3. Why would it be better to use larger samples of beans to find results?

4. In your own words, explain the difference between Average Atomic Mass (as found on your periodic table) and Mass Number.

5. Why are these beans not considered "true isotopes?" Your answer should include the definition of an isotope.

6. Calculate the average atomic mass of magnesium. Use the example on the first page as a guide.

Isotope	Percent Abundance
Mg-24	79%
Mg-25	10%
Mg-26	11%

Half-Life Lab – Teacher Notes

Radioactive isotopes of elements naturally decay to more stable isotopes of other elements. It takes a certain amount of time for half of the material to decay. This is called half-life. This lab simulates half-life using the basic probability of flipping a coin (or M&M candy). Before this activity, I have students work on a research assignment for two periods.

Time: One class period

Materials per group

- one large plastic cup
- small cup with M&Ms (or pennies)
- gloves (if working with M&Ms)
- blank sheet of paper (if working with M&Ms)

Notes

Buying enough M&Ms can be expensive, especially since they can't be reused. An alternative is to use pennies. I give out about 100 to each group to keep them counting. When working with M&Ms, I fill a small bathroom cup with the candy, one for each person. They get into groups and pool together all the M&Ms for the lab, then dole them back out at the end. I have the students run two trials, but based on time, you can have them run one.

I'm including both versions of the lab sheet here, one for M&Ms and one for pennies.

To the class

Today we're dealing with a radioactive isotope that will undergo decay to a safe element. [If M&Ms, I don't tell them what they are yet. I also mention they will need to wear gloves.] You must first count all the isotopes you have and record it under half-life zero. You will next place the isotopes into the large cup and shake it. Pour it out onto the table [if M&Ms, onto the blank sheet of paper]. Some of the isotopes will have decayed to a safe state – they will be upside down [if M&Ms: not showing a marking, if pennies, tails side up]. Remove these for now [if M&Ms: but don't eat them]. We'll be running a second trial. Count the number of isotopes that are face-up and record it. Put these back in the cup, shake, repeat. Do this until there are no more radioactive isotopes [face-up] left.

Discussion

This lab generally produces a nice half-life curve. We run the second trial to help this along, but you can eye the data and instead choose a group to only plot one trial instead of the average. Most students get the third analysis question wrong, thinking maybe they can run ten more trials, or so on. They should only be able to run one more trial, because this is half-life. If they went up, it would double once. Their graph should show this.

Name Period Date Teacher

Half-Life and Nuclear Reactions Research

Use the internet and your knowledge of science to answer the following. Do not copy and paste answers.

1. What is radioactivity?

2. When is an isotope considered stable?

3. How are radioactive isotopes harmful or dangerous?

4. How can radioactive isotopes be useful?

5. For each isotope shown below, describe what it is used for.

    a. iodine-131

    b. cobalt-60

    c. carbon-14/carbon-12 ratio

    d. uranium-238/lead-206 ratio

6. What is half-life?

7. What is the half-life of the following isotopes?

gold-198		carbon-14		calcium-37	
francium-220		plutonium-239		uranium-238	

8. What are the following particles also known as?

alpha particle		beta particle	
positron		deuterium	

9. What is a decay mode?

10. What is the decay mode for the following isotopes?

gold-198		carbon-14		calcium-37	
francium-220		plutonium-239		uranium-238	

11. What is a fission reaction?

12. Show an example of a fission reaction using chemical symbols (this may be pasted in).

13. Where are three fission reactors located across the world?

14. What is a fusion reaction?

15. Show an example of a fusion reaction using chemical symbols (this may be pasted in).

16. Where is the largest fusion reactor on Earth located?

17. What type of reaction powers the sun: fission or fusion?

18. What are three famous nuclear reactors and why are they famous?

Nuclear reactor	**Why it's famous**

19. What are three main parts of a nuclear reactor and what do they do?

Nuclear reactor part	**What it does**

20. What are some dangers of nuclear power? List three.

21. What are some benefits of nuclear power? List three.

20. What is a nuclear meltdown?

21. What is a nuclear chain reaction?

22. What are the components of a nuclear bomb?

23. Briefly describe the Manhattan Project (not the movie). What was its goal?

24. When were nuclear weapons first used in warfare?

25. What are some towns or cities that have been destroyed by nuclear weapons?

26. Where does nuclear weapon testing take place inside the US? Where outside the US?

27. Who are some inventors of the atomic bomb?

28. Briefly describe the nuclear arms race during the Cold War.

29. Opinion Essay. Should countries continue to develop nuclear weapons? (Write at least 100 words.)

30. Opinion essay. Should countries continue to create nuclear power plants and use nuclear energy? (Write at least 100 words.)

Names Period Date Teacher

Half-Life Lab – With M&Ms

Purpose: Radioactive materials decay at a rate known as half-life. Half-life is defined as the amount of time that is required for a material to decay by half its amount. You will simulate radioactivity in nature.

Materials: Clean sheet of paper or large plate, small plates (one for each member), cup of M&Ms, gloves

Safety: The materials are considered radioactive and dangerous. They must be handled with gloves. Do not eat any M&Ms until the experiment is completely over (2 trials) and they have decayed to a safe isotope.

Procedure:

1) Count the number of M&Ms you have and record the number as Half-Life 0 in Trial 1.
2) Place all the M&Ms in the cup.
3) Cover the cup and shake it multiple times.
4) Spill the M&Ms onto a clean sheet of paper or plate.
5) Remove all of the M&Ms that are not facing up. These M&Ms have decayed into a safe, edible isotope.
6) Count the remaining M&Ms that are facing up. Record the number in the data table in Trial 1.
7) Repeat steps 2 – 6 until all M&Ms are gone.
8) Repeat the experiment for Trial 2, then average the values together
9) Once the experiment is over, plot your data (using the average) and connect the points with a curved line.
10) Eat your safe isotopes, discard all materials, and answer the questions.

Half Life	Trial 1 # M&Ms	Trial 2 # M&Ms	Avg
0			
1			
2			
3			
4			
5			
6			
7			
8			
9			
10			
11			
12			
13			
14			
15			

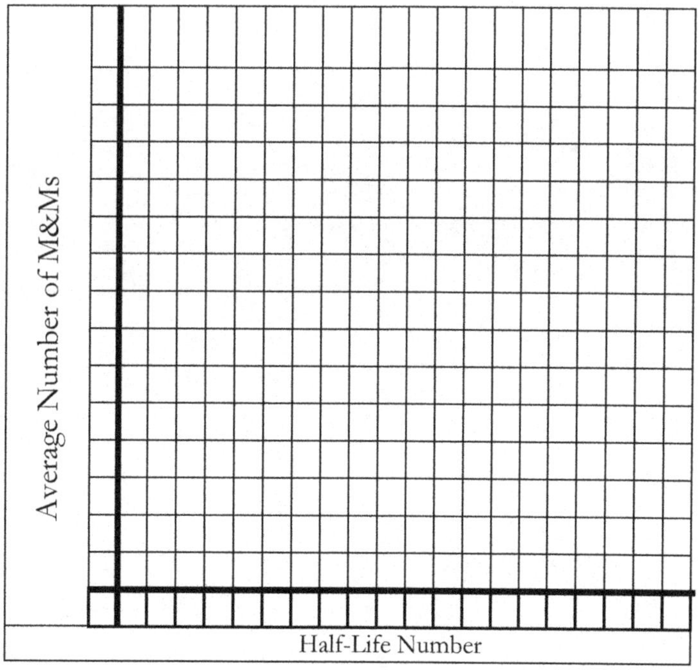

Analysis:

1. As the number of trials increases, what happens to the number of M&Ms facing up? ______________
2. What would you expect the number of M&Ms to be at half-life 3.5? ______________________
3. If you started with twice as many M&Ms, how many more readings would you have been able to take?

Names Period Date Teacher

Half-Life Lab – With Pennies

Purpose: Radioactive materials decay at a rate known as half-life. Half-life is defined as the amount of time that is required for a material to decay by half its amount. You will simulate radioactivity in nature.

Materials: Cup of pennies, plate (optional)

Procedure:

1) Count the number of pennies you have and record the number as Half-Life 0 in Trial 1.
2) Place all the pennies in the cup.
3) Cover the cup and shake it multiple times.
4) Spill the pennies onto the table or a plate.
5) Remove all of the pennies that are not facing up. These pennies have decayed into a safe isotope.
6) Count the remaining pennies that are facing up. Record the number in the data table in Trial 1.
7) Repeat steps 2 – 6 until all pennies are gone.
8) Repeat the experiment for Trial 2, then average the values together
9) Once the experiment is over, plot your data (using the average) and connect the points with a curved line.
10) Return the pennies and answer the questions.

Half Life	Trial 1 # Pennies	Trial 2 # Pennies	Avg
0			
1			
2			
3			
4			
5			
6			
7			
8			
9			
10			
11			
12			
13			
14			
15			

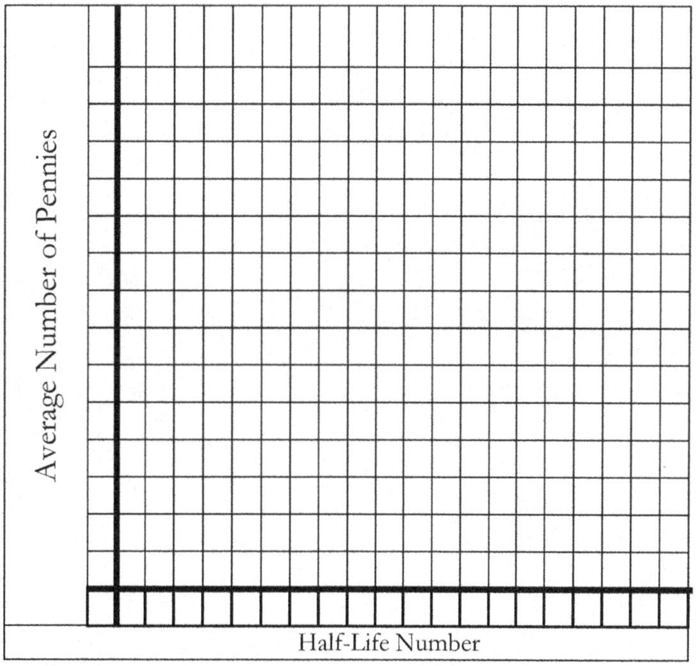

Analysis:

1. As the number of trials increases, what happens to the number of pennies facing up? ______________
2. What would you expect the number of pennies to be at half-life 3.5? __________________________
3. If you started with twice as many pennies, how many more readings would you have been able to take?

CHAPTER 4

Chemical Reactions

Chemical reactions are at the heart of chemistry. It's what makes the subject so much fun. There are several types of chemical reactions, and they can manifest themselves in many ways. This section deals with several reactions that are sure to entertain.

Basic Types of Chemical Reactions

In a ***synthesis reaction***, two (or more) substances are brought together and create one type of product. An example of this is the formation of water from its component elements.

$$2H_2 + 2O_2 \rightarrow H_2O$$

A ***decomposition reaction*** is the opposite of a synthesis reaction. In this, a compound is broken up into smaller parts. We can also show this with water, this time breaking into its elements.

$$H_2O \rightarrow 2H_2 + 2O_2$$

Single replacement reactions occur when an element replaces another element in a compound.

$$2NaBr + Cl_2 \rightarrow 2NaCl + Br_2$$

A ***double replacement reaction*** occurs when two ions switch places between two compounds. Neutralization reactions between acids and bases are double replacement reactions.

$$NaOH + HCl \rightarrow NaCl + H_2O$$

Organic Reactions

Organic compounds all require carbon, and many contain *functional groups* that give them specific properties, such as -OH for alcohols and -COOH for organic acids. These compounds have their own set of reactions.

Addition: This is like a synthesis reaction
Substitution: This works like a double replacement reaction
Polymerization: Smaller molecules (monomers) combine into larger molecules (polymers)
Esterification: An organic acid and an alcohol combine to form an ester, a scented compound
Fermentation: Converts a carbohydrate into an alcohol, such as in the presence of yeast
Saponification: Animal or vegetable fat is converted into soap in the presence of an alkali, such as NaOH
Combustion: The burning of oxygen releases carbon dioxide and water

Neutralization – Teacher Notes

Neutralization reactions occur when an acid and a base react. This produces a type of salt and water, as well as an occasional additional side product, such as CO_2. This lab tests various liquids using sodium bicarbonate to see if a reaction occurs, implying it is acidic.

$$\text{Acid} + \text{Base} \rightarrow \text{Salt} + \text{Water} (+ CO_2)$$

<u>Time</u>: One class period

<u>Lab safety</u>
There will be normal food items (liquids) in the room. It is imperative to remind students they are not to sample any of them.

<u>Materials</u>

- Up to 20 different liquids to test
    - 75 mL of each should be more than enough for one class
        - It's 5 mL per trial (times the number of groups/pairs)
    - Water, soda, vinegar, sports drinks, tabasco sauce, lemonade, grape juice, milk, orange juice, etc.
    - A beaker, 10 mL graduated cylinder, and dropper for each liquid sample
- Three to four balances at one station
    - sodium bicarbonate – 250 grams total should be enough for one class
        - 1 gram per trial (times 20 trials per group)
    - weighing tray at each balance
- One cup or 100 mL beaker per group
- One stopwatch per group

<u>Notes</u>
I usually have students work in pairs because there are so many stations to try. They can work in any order, and this freedom works well, I find. Obtaining 20 liquids can be pricey unless you can scrounge around and get donations from other teachers. You don't need a lot of each liquid. You can also skip the balances and have students use a small scoop of sodium bicarbonate instead.

Not all acids will produce bubbles with sodium bicarbonate, so they may find a few false negatives.

<u>To the class</u>

Today we're carrying out neutralization reactions. [Explain what a neutralization reaction is.] We're testing this out using sodium bicarbonate. You're going to go around the room to see if each liquid reacts with sodium bicarbonate. If it does, we will assume it's an acid. If it doesn't, we will assume it's a base. Your first step is to take a reaction cup [or beaker] and measure out 1 gram of sodium bicarbonate into it. [Either use a balance or just a small scoop.] Then go to a station and measure out 5 mL into the graduated cylinder. Get your stopwatch ready, then pour the liquid into your reaction cup and time it to see how long the reaction occurs. Would a fast reaction mean the acid is stronger or weaker? [stronger] Keep that in mind. Some liquids may not react at all. Once you have your data, rinse out your reaction cup and start again. Don't dry the cup in between trials. It's not necessary. Please make sure you don't walk away with the dropper at the station. Keep your stopwatch with you. Also, you must not taste any of the materials at the stations. You can move to any free station when you're ready.

Discussion

On a second day, have students share their results with the class. Make note on the board which liquids they thought ranked 1, 2, or 3 for strongest acid. Then make a list for 4, 5, 6, and so on. At some point, the students will likely have struggled to rank the liquids because they did not react. Cluster these together.

Here you can discuss the pH scale, which runs from 1 (strong acid) to 14 (strong base) with 7 being neutral (such as water). Relate the pH scale to the findings. The stronger acids would be closer to 1 in the ranking, though you won't actually have any liquids with a pH of 1 in the lab. This is just to get a sense of how the scale works.

Next, discuss how there are other, more accurate ways to determine if something is an acid or a base. You can use Litmus paper, which turns red in an acid and blue in a base. There is also universal pH paper that can be used to give an approximate pH. If you have these materials, you could potentially run a follow-up lab to test the liquids again to see how the results match.

The neutralization reaction between vinegar (acetic acid) and baking soda (sodium bicarbonate) is as follows.

$$NaHCO_3 + HC_2H_3O_2 \rightarrow NaC_2H_3O_2 + H_2CO_3$$

The reaction produces carbonic acid (H_2CO_3), but it's very unstable, so it breaks apart into water and carbon dioxide. It's the release of CO_2 that makes the bubbles. The $NaC_2H_3O_2$ (the salt in this reaction, sodium acetate) breaks apart in water to form Na^+ and $C_2H_3O_2^-$ ions.

That means this is the actual final result of the neutralization.

$$NaHCO_3(s) + HC_2H_3O_2(l) \rightarrow Na^+(aq) + C_2H_3O_2^-(aq) + H_2O(l) + CO_2(g)$$

Note: (s) = solid, (l) = liquid, (g) = gas, (aq) = aqueous or dissolved in water

Names Period Date Teacher

Neutralization Lab

Problem/Research/Hypothesis. A neutralization reaction occurs when an acid and a base interact to produce a salt and water. Some neutralization reactions also produce another byproduct, such as CO_2. In this experiment, you will examine the results of a known neutralization reaction, and based on the result, you will then try to determine whether other chemicals are acids or not. You will also attempt to rank the unknown liquids in order of most to least acidity.

$$\text{Acid} + \text{Base} \rightarrow \text{Salt} + \text{Water} \ (+\ CO_2)$$

Procedure

1) Rinse out the reaction beaker or cup with water. Do not dry.
2) Measure out 1 gram of sodium bicarbonate.
3) Place the sodium bicarbonate into the reaction beaker or cup.
4) Go to an open "unknown liquid" station.
5) Use the beaker and the pipette to measure 5 mL of the liquid into the graduated cylinder.
6) Once you have measured out the 5 mL of liquid, prepare the stopwatch.
7) Pour the liquid into the reaction beaker or cup containing the sodium bicarbonate.
8) Measure the time it takes for the reaction to complete and record it in the data table.
   a) If nothing happens, put a dash (-) for time
9) Make note of what you observed during the reaction (i.e.: bubbling, color change, etc.)
10) If no reaction appears to take place, make note of that in the data table.
11) Based on your observations of the known liquid, guess if the unknown liquid is an acid or not.
12) Repeat these steps until time runs out.

Analysis and Conclusion

1. Based on your data, which liquids do you think are acids? List all.

2. If you are doing a lab and you accidentally spill some acid on yourself, what could you do? Base your answer on what you learned in this lab.

3. Examine your results in the data table. Based on your findings, determine which liquids you feel are stronger acids. Rank all of the liquids in the chart based on your observations. For the strongest acid, assign it the number 1, etc. If you think that two liquids are the same strength, then mark them with the same rank number.

4. Sometimes when you eat your stomach produces too much hydrochloric acid (HCl) and you feel unwell. For this you can take something like Alka-Seltzer. Based on your findings in this lab, describe what Alka-Seltzer is and what it is designed to do.

Sample #	Name of Liquid	Reaction Time (sec)	Reaction Observation (what you saw)	Do You Think It Is an Acid?	Rank (1 = strongest)
1					
2					
3					
4					
5					
6					
7					
8					
9					
10					
11					
12					
13					
14					
15					
16					
17					
18					
19					
20					

Red Cabbage Indicator – Teacher Notes

The pH scale helps us differentiate the strengths of acids and bases. Juice from the leaves of red cabbage provides a nice visual of materials at several pH levels. Note that this lab uses a hot plate with boiling water, then time is needed to cool the water down. It might be useful to (1) break this activity into two periods, or (2) prepare the indicator ahead of time. When using hot plates, it is imperative to wear goggles at all times.

Time: Two class periods (if the students are making the indicator, otherwise one period)

Lab safety

For this lab, you will need to use hot plates. Therefore, lab safety must be made a serious concern. Students are not to touch the beaker or the hot plate surface once the hot plates are turned on. Goggles must also be worn at all times.

Materials per group

- Hot plate
- Two 400 mL beakers
- One leaf of red cabbage (one head of cabbage will last a few classes)
- Paper plate to hold the red cabbage shreds
- Strainer or wire screen
- Glass stirring rod
- Thermometer
- Beaker tongs
- 12 test tubes
- Test tube rack
- Small funnel

Materials to test

- Acids: powdered toilet clearer (strong), vinegar, cream of tartar (weak)
- Neutral: water
- Bases: dishwasher powder (strong), milk of magnesia, sodium bicarbonate (weak)
- Unknowns: borax, lemonade, citrus soda, white grape juice, sugar
- One dropper per liquid and one metal scoop per solid
- 50 mL beaker (or bathroom cup) per each test material

Notes

If you make the juice indicator ahead of time, you will only need the test tubes and rack, the solution (indicator), and the materials to test. The funnel is helpful when pouring the indicator into the test tubes, but isn't required. The unknowns are suggestions. Only their pH is unknown for this lab; you can tell the students what the materials are. I usually leave all the containers out for reference. I set up a station on the side of the room with all the test materials. Students visit the station to gather whatever sample they need. They won't need much; just a few squirts of liquids and a single scoop of solids.

Discussion

pH is a measure of hydrogen ions (H^+) or hydronium ions (H_3O^+) in a solution. Each value on the pH scale represents a concentration. A pH of 5 has an H^+ ion concentration of 1×10^{-5}. A pH of 8 has a concentration of 1×10^{-8}. Each 1 value change in pH relates to a 10-fold change in concentration. It's an inverse relationship; as one value goes up, the other goes down. So, if pH decreases from 5 to 4, the concentration increases by a factor of 10. If pH decreases from 5 to 3, that's 2 changes in pH, so the concentration increases by 10 x 10 or 100. If pH increases from 7 to 10, that 3 changes of pH, so the concentration decreases by 10 x 10 x 10 or 1000.

In our lab, we established the relative pHs of several materials, then used that to identify whether other materials were strong or weak acids or bases. Red cabbage juice acts like a universal indicator, so it can respond to materials at a wide range of pH values. It contains a compound, anthocyanin, that responds to the acidity or alkalinity of materials. A reddish/purplish color relates to higher acidity. Greenish/yellowish is more basic (or higher alkalinity).

Names Period Date Teacher

Red Cabbage Indicator Lab

Purpose: Use a red cabbage indicator to establish a pH scale, then use it to identify unknown materials.

Safety: Goggles must be worn at all times. Be careful of hot liquids. Do not touch test materials with bare hands.

Materials: red cabbage leaf, two 400 mL beakers, hot plate, wire gauze, 12 test tubes, paper plate, glass rod, thermometer, small funnel

Chemicals to Test: (there may be some substitutions)

powdered toilet cleaner (SA)	vinegar (A)	cream of tartar (WA)	water (N)
sodium bicarbonate (WB)	milk of magnesia (B)	dishwasher powder (SB)	borax
lemonade	Sprite	white grape juice	sugar

NOTE: SA = strong acid, A = acid, WA = weak acid, N = neutral, WB = weak base, B = base, SB = strong base

Indicator Preparation Procedure

1) Add about 275 mL water to a 400 mL beaker and set it on the hot plate
2) Heat the water to boiling then turn off the heat
3) While waiting for the water to boil, shred the cabbage leaves into small pieces
    a) Place on the paper plate
4) Once the water boils, add the shredded leaves to the water and stir until there is a deep color in the water
5) Let the mixture cool until it is below 60 °C (about 15 minutes)
6) Using the beaker tongs, strain the mixture into the other beaker
7) Discard the cabbage bits in a garbage can, not down the sink; clean the beaker

Testing Procedure

1) Fill the test tubes less than halfway full with the red cabbage juice (indicator)
   a) The exact amount does not matter
2) Test the known samples by adding each to a separate test tube containing the indicator
   a) You can do this in any order; make note of the material and its relative acidity
   b) If the sample is a liquid, add a couple squirts to a test tube. Use one scoop if it's a solid.
3) Make note of any color change
4) Plot the colors on the given pH scale in order from most acidic (1) to most basic (14)
5) Obtain the unknown samples one at a time and determine their approximate pHs
6) Rinse all materials with plenty of water

Known/Unknown	Material	Relative Acidity	Color	Approx. pH
Known				
Known				
Known				
Known				
Known				
Known				
Known				
Unknown 1				
Unknown 2				
Unknown 3				
Unknown 4				
Unknown 5				

Estimated pH Scale. Plot the solution colors here in order of acidity using the known materials, then use the scale to determine the acidities of the unknowns. Remember, 7 is neutral.

1 2 3 4 5 6 7 8 9 10 11 12 13 14

Analysis

1. What were the three strongest acids among all the samples tested (including the unknowns)?

2. What were the three strongest bases among all the samples tested (including the unknowns)?

3. What are two compounds you could have potentially mixed together to make a neutral solution?

4. You test an unknown material and it creates a reddish-pink color in the cabbage solution. What could this compound be, according to your data?

5. You have a compound with a pH of 10. What color would you expect the cabbage solution to turn?

Polymerization – Teacher Notes

When smaller molecules, known as monomers, join together in a long chain, called a polymer, we have a process known as polymerization. A favorite activity to demonstrate this process is making slime. There are several methods of making slime, but this one sticks to the basics with no bells or whistles. It highlights just the polymerization reaction.

Time: One class period

Materials per group

- 25 mL polyvinyl alcohol (Elmer's glue)
- 25 g sodium borate (Borax laundry detergent)
- 100 mL graduated cylinder
- Two large plastic cups
- Two spoons
- Dropper
- Tray (optional)
- Gloves (optional)
- Food coloring (optional)
- Small sealable bag (optional)

Notes

It is possible for students to overshoot the reaction and not create slime, so it's important they add the Borax solution slowly to the glue mixture. (Of course, more glue could be added, but I try to avoid that.) When the reaction does make the slime, it might be "too slimy" and messy. They can add some dry Borax to this, a few crystals at a time and work it in to fix it. Depending on your school and the maturity of the students, you can offer a small baggie for them to take the slime with them. Food coloring makes a bit of a mess too, but the students like to add the colors. I also offer gloves for those who are reluctant to touch the slime directly. It's nontoxic, just goopy. When properly made, the slime will form into a jiggly bouncy ball. It fractures and tears pretty easily, but it can be mashed back together. Often, I will allow students to each create their own ball of slime. They share the Borax solution in groups. Generally, I will pour out the glue so it's a more controlled amount.

Other variations include the addition of hand lotion or shaving cream. These give different consistencies (and scents). You could adapt this lesson to include these ingredients to test which slime is the best.

To the class

Today we're conducting an experiment to produce a material known as a polymer. This can get a little messy, so we need to keep things as neat as possible. [If you have trays, all work should be done on them.] You will start by making two solutions. In the first, you will dilute some polyvinyl alcohol with an equal amount of water. In the second, you will create a saturated solution of sodium borate. [Saturated means no more can dissolve, so some powder will sit at the bottom of the cup.] You will then ***slowly*** add the second solution to the first, stirring along the way. You will use a dropper to transfer the solution. If you add too much, it will ruin the reaction, so you need to be careful. For cleanup, all materials will be discarded [unless you're letting them keep the polymer in a baggie].

Discussion

Polyvinyl alcohol [$C_{12}H_{26}O_6$] is a large molecule with lots of -OH groups, which allows it to make hydrogen bonds, and therefore it dissolves well in water.

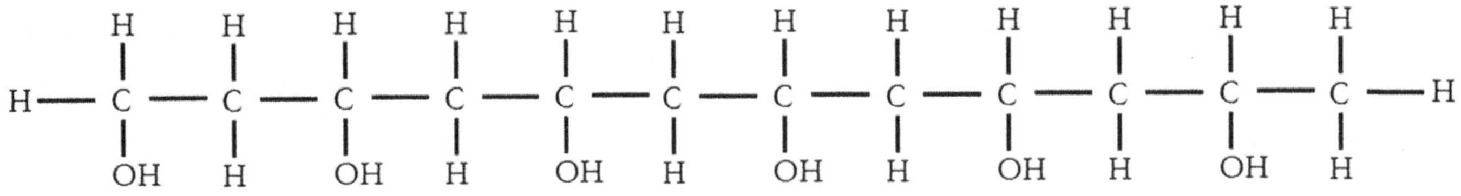

The borate ion [$B(OH)_4^-$] is tetrahedral in shape. It looks sort of like a pyramid. Here's a flat view of the ion.

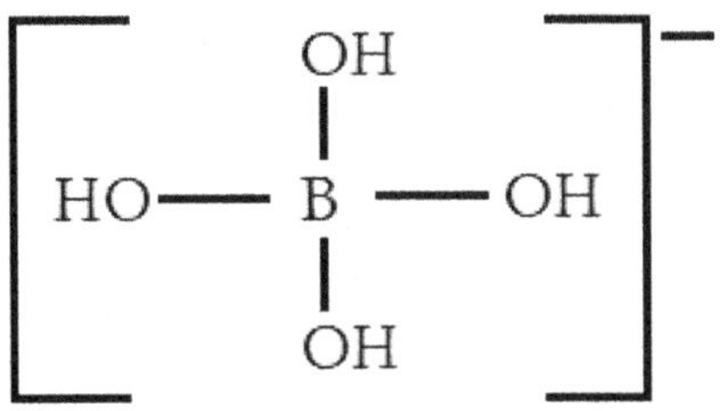

Hydrogen bonds are strong ***intermolecular forces*** that exist between two molecules. The oxygen atoms from the borate ion move close to the hydrogen atoms of the -OH groups of the polyvinyl alcohol. This causes the long chains to twist and bend. It also links separate chains together, making a longer chain known as a polymer.

Names Period Date Teacher

Polymerization Lab

Polymers are long chains of molecules made up of smaller molecules, known as monomers. An easy polymer can be made with two household chemicals, polyvinyl alcohol and sodium borate. You must take care, however, not to mix too much of the sodium borate, since it will ruin the composition of the polymer. Carefully follow the procedure below.

Do not eat or taste any of the chemicals. Be sure to wash your hands at the end of the lab.

Write the household names of the following chemicals:

Polyvinyl alcohol ______________________ Sodium borate ________________________

Materials: 25 mL polyvinyl alcohol, 25 g sodium borate, 100 mL graduated cylinder, 2 plastic cups, 2 spoons, dropper, plastic bag (optional)

Procedure

Mixture A

1) Measure out polyvinyl alcohol into a plastic cup up to the first ring in the cup (about 25 mL)
   a) Do NOT pour polyvinyl alcohol into a graduated cylinder
2) Measure out an equal amount of water (about 25 mL) and add it to the polyvinyl alcohol
3) Mix with a spoon

Mixture B

1) In a separate cup, create a saturated solution of sodium borate
2) Add 75 mL water to the cup and mix in a scoop of sodium borate. Stir with the second spoon. Add more sodium borate as needed until no more dissolves.

Mixture C

1) You will create the third mixture by **slowly** adding mixture B into mixture A.
2) You must carry this out slowly and write your observations with each addition.
3) Make your initial observation in a data chart before mixing A and B.
4) Use a pipette (dropper) to withdraw some of Mixture B (saturated sodium borate)
5) Add a squirt (roughly 10 drops) to the plastic cup with the polyvinyl alcohol solution.
6) Stir vigorously without spilling. Once the reaction begins, stir for at least a minute or two.
7) Record your observations on the back.
8) Repeat steps 4-7 until all of Mixture A has reacted. You will have to use your best judgment. Every Mixture A requires a slightly different amount of Mixture B.
9) Remove the polymer from the plastic cup and make new observations. Remember to keep the polymer in your workspace.
10) If needed, you may add some sodium borate crystals to the polymer and work them in to toughen it up.
11) Discard Mixture B with plenty of water.
12) Plastic cups and wood stirring rods must be discarded.

Total # Drops Added	Consistency	Texture	Observations

Analysis and Conclusion

1. What happened as you added more Borax to the glue?

2. Roughly how many drops of Borax solution did you need for the reaction to be complete?

3. Why do you think water was added to Mixture A?

4. What are some possible uses of this substance?

5. What would you change in this experiment to make it better?

Clock Reaction – Teacher Notes

This is one my all-time favorite labs… You mix two solutions together and nothing happens. Then a minute later, *voila!*, it turns blue.

Time: One class period

Lab safety

Goggles must be worn. Students will be using 2% iodine solution and 3% hydrogen peroxide in this lab. Both can damage clothing, and must be kept out of eyes.

Materials per group

- 25 mL 2% iodine
- One 1000 mg Vitamin C tablet
- 15 mL liquid starch
- 75 mL 3% hydrogen peroxide
- Stopwatch
- Teaspoon
- Mortar and pestle (one can be shared by the class if needed)
- Three stirring sticks
- Two 100 mL beakers
- Two 250 mL beakers
- 100 mL graduated cylinder

To the class

In this lab, we're making use of iodine and starch to produce a clock reaction. When mixed, it takes time for these two chemicals to interact, and once they do, the solution changes color. You'll be testing the solution at different concentrations to see the effect on how much time it takes for the color change to occur.

We're using an iodine solution, which can easily and permanently stain clothing. You will also be using hydrogen peroxide, which can have a similar effect. It is important you handle these carefully. Because of these chemicals, goggles must be worn at all times.

This procedure is a bit detailed, and you must take care to follow it properly. You will be creating four solutions, one of which is your final reaction mixture. Adding materials in the wrong order will ruin the experiment, so take care to follow each step exactly. You will repeat this experiment using different concentrations of one solution. Between trials, do not clean out beaker A, as you will need it throughout.

Discussion

The first thing that happens is hydrogen peroxide reacts with iodide ions to create elemental iodine. When starch and elemental iodine react, they produce a dark blue color. This is one chemical reaction that takes place in this experiment, and it reacts fast. However, the vitamin C inhibits the reaction by soaking up elemental iodine and preventing it from reacting with the starch. That is, until the vitamin C is consumed, after which the solution turns blue. The reason for the delay is the consumption of the vitamin C.

Names Period Date Teacher

Clock Reaction Lab

Purpose: To explore an iodine clock reaction

Safety: BE CAREFUL WITH IODINE. IT WILL STAIN CLOTHING. Goggles must be worn.

Materials: 25 mL 2% iodine, 1 1000-mg Vitamin C tablet, 15 mL liquid starch, 75 mL 3% hydrogen peroxide, stopwatch, teaspoon, mortar and pestle, 3 stirring sticks, 2 100-mL beakers, 2 250-mL beaker, graduated cylinder

Note: Use separate stirring sticks for each solution

Procedure 1: 100% Concentration

1) Label the 100 mL beakers A and B; label the 250 mL beakers C and D
2) SOLUTION A:
   a) Add 60 mL warm water to Beaker A
   b) Crush the Vitamin C tablet and add to Beaker A; stir with a stirring stick
3) SOLUTION B:
   a) Add 60 mL warm water to Beaker B
   b) Add 1 teaspoon (5 mL) of the solution from Beaker A into Beaker B
   c) Rinse off the teaspoon
   d) Add 0.9 teaspoon (4 mL) iodine solution to Beaker B; stir with a stirring stick
   e) Rinse off the teaspoon
   f) NOTE: If solution B is dark brown, add more of solution A
4) SOLUTION C:
   a) Add 60 mL warm water to Beaker C
   b) Add 1 tablespoon (3 teaspoons or 15 mL) hydrogen peroxide to Beaker C
   c) Rinse off the teaspoon
5) Add ½ teaspoon (2.5 mL) liquid starch to Beaker C; stir with a stirring stick
6) Rinse off the teaspoon
7) Pour solution B into the reaction beaker D
8) Prepare the stopwatch
9) Add solution C to reaction beaker D and start the stopwatch
10) Stop the timer when the solution stops changing color; record the time under "Concentration 100%"
11) Wash out Beakers B and C and the reaction beaker D
    a) **<u>Do not</u>** wash out Beaker A at this time

Procedure 2: 66% Concentration

1) Recreate Solution B in Beaker B
2) Recreate Solution C in Beaker C using 90 mL water (instead of 60 mL)
3) Repeat the reaction from Procedure 1 (Steps 5 through 10)

Procedure 3: 50% Concentration

1) Recreate Solution B in Beaker B
2) Recreate Solution C in Beaker C using 120 mL water (instead of 60 mL)
3) Repeat the reaction from Procedure 1 (Steps 5 through 10)

Procedure 4: 33% Concentration (If time permits)

1) Recreate Solution B in Beaker B
2) Recreate Solution C in Beaker C using 180 mL water (instead of 60 mL)
3) Repeat the reaction from Procedure 1 (Steps 5 through 10)

Clean up:

1) Clean out all beakers
2) Make sure chemical bottles are sealed
3) Discard the stirring sticks and spoon
4) Return lab materials
5) Wash your hands

Data Table

Concentration	Reaction Time (s) **
100%	
66%	
50%	
33%	

** Convert minutes to seconds as needed (i.e: 2:14 → 134 sec)

Analysis

1. What was the general time trend as the concentration decreased?

2. Describe what you would need to do to reduce the reaction time.

3. What would you expect the reaction time to be in a 10% solution?

Elephant Toothpaste – Teacher Notes

Elephant toothpaste is a fun and messy lab. Mixing several chemicals together in a large cylinder or bottle, a massive eruption occurs with rising foam. Kids love it, but they need to be careful of the hydrogen peroxide.

Time: One class period

Materials per group

- 150 mL 6% hydrogen peroxide
- 1 L graduated cylinder
- Tall-walled tray (i.e.: aluminum roasting pan)
- Thermometer
- 2 packets or 14.0 grams yeast
- 100 mL warm water (25°C)
- 100 mL beaker
- 15 mL dish soap
- Goggles
- Stirring stick
- Stopwatch
- Optional (but recommended)
    - Ring stand with a thermometer clamp

Notes

If you don't have access to 1 L graduated cylinders, you can use 2 L bottles with the tops cut off. Be sure to smooth the jagged edges or cover them with duct tape. If you use the bottles with the spouts attached, you're asking for a real mess. The lab calls for the students to measure the temperature of the foam, which can be tricky to do before it cools. The thermometer should be held horizontally into the foam so the students don't get their hands in the heat directly. A ring stand with a thermometer clamp helps ease this process. For the 3% concentration in a 1 L cylinder, the foam will reach up to the top. With the 6%, it will undoubtedly spill over. It's difficult to estimate the amount of foam, but the conversation with the students is always worth it as they try to figure out a method.

To the class

The circus is in town, and you're charged with taking care of an elephant. In particular, you need to make sure its teeth are clean, so in this lab, we're making elephant toothpaste. This lab can be a bit messy, so we're using these trays to help keep things clean. We're also using a stronger version of hydrogen peroxide than you probably have at home, so we need to be careful not to get it on our skin or clothes. We're going to wear goggles today to protect our eyes. One important thing is to make sure you get warm water that's about 25°C. Too low, and the yeast won't activate. Too high, and it will kill the yeast. I suggest going a little warmer and letting it cool before you use it.

Discussion

When yeast and hydrogen peroxide (H_2O_2) mix, a chemical in the yeast breaks down the peroxide and releases oxygen gas and heat. By adding dish soap, the oxygen gas grabs on and makes bigger bubbles, which creates the fountain of foam we see.

Names Period Date Teacher

Elephant Toothpaste Lab

Purpose: To explore a chemical reaction

Materials: 150 mL 6% hydrogen peroxide, 1 L graduated cylinder, tray, thermometer, 1 packet or 14.0 grams yeast, 100 mL warm water (25°C), 100 mL beaker, 15 mL dish soap, stopwatch, stirring stick

Safety: Do not touch the hydrogen peroxide with your hands or get it on your clothes

Procedure

Part 1: 3% hydrogen peroxide

1) Add 50 mL water to the 1 L graduated cylinder
2) Add 50 mL 6% hydrogen peroxide to the 1 L graduated cylinder
3) Add one "squirt" (about 5 mL) dish soap [don't go crazy]
4) Gently swirl the 1L graduated cylinder to mix the contents
5) ***In a separate beaker***, add 50 mL warm water (about 25°C) and 7.0 g of yeast (1 packet)
   a) Stir for about 30 seconds
6) Record the temperature of the yeast mixture
7) Prepare a stopwatch and have the thermometer ready
8) Add the yeast mixture to the bottle swiftly and start the timer
9) Record the amount of time the reaction runs for
10) Also, record the temperature of the foam before it cools
11) Estimate the amount of foam created
12) To clean up efficiently, add some water to the large graduated cylinder, shake and swirl it around

Part 2: 6% hydrogen peroxide

1) Add 100 mL 6% hydrogen peroxide to the clean bottle
2) Go to step 3 in Part 1 and follow the rest of the procedure

Hydrogen Peroxide Strength	Initial Temperature	Final Temperature	Change in Temperature $T_{final} - T_{initial}$	Reaction Time	Amount of Foam
3%					
6%					

Analysis

1. Was the reaction endothermic or exothermic? Endothermic means the solution got colder by absorbing energy. Exothermic means it heated up by releasing energy.

2. Which reaction ran for the longest? Why do you think that happened?

3. Which concentration of hydrogen peroxide produced the most heat? Why do you think that happened?

4. Which concentration of hydrogen peroxide produced the most foam?

5. What would you expect to happen if 12% hydrogen peroxide was used?

6. Using 6% concentration hydrogen peroxide, how could you produce twice as much foam? Be specific.

7. What was the purpose of adding dish soap?

8. What would have happened if a bottle or flask was used that had a narrower neck?

Reaction Kinetics – Teacher Notes

There are several factors that affect the rate of a chemical reaction, including temperature, concentration, nature of reactants, surface area, and presence of a catalyst. This lab explores the first four of these conditions. Note that this lab utilizes a large amount of vinegar. If your school has a science supply budget, it may be worth utilizing it to obtain the vinegar. I run this lab in four stations, usually with two setups per station to keep groups smaller. This, of course, doubles the amount of vinegar and other supplies. The vinegar stations utilize two trials per attempt, so you could reduce your vinegar use by half by running only a single trial.

<u>Time</u>: Two class periods

<u>Materials per station rotation</u>

- ***Station 1 – Temperature***
    - 300 mL vinegar
    - 30 g sodium bicarbonate
    - 100 mL beaker
    - Metal scoop
    - Thermometer
    - Stopwatch
    - Ice cubes (just a few)
    - Warm vinegar
    - Weighing tray and balance (optional)
- ***Station 2 – Concentration***
    - 300 mL vinegar
    - 30 g sodium bicarbonate
    - 250 mL beaker
    - 100 mL graduated cylinder
    - Metal scoop
    - Weighing tray and balance (optional)
- ***Station 3 – Surface Area***
    - 6 antacid tablets
    - Mortar and pestle
    - 250 mL beaker
    - 100 mL graduated cylinder
    - Weighing tray (to hold tablet pieces and powder)
- ***Station 4 – Nature of Reactants***
    - 100 mL vinegar
    - 100 mL lemon juice or strong lemonade
    - 30 g sodium bicarbonate
    - 250 mL beaker
    - 50 mL graduated cylinder
    - Stopwatch
    - Metal scoop
    - Weighing tray and balance (optional)

Yes, that's 700 mL of vinegar per two trials per group!

Notes

To save time during the lab, I generally have the students measure a heavy scoop of sodium bicarbonate rather than measure 5 grams with the balance. If you don't have hot water available, you could use a hot plate at a low temperature to keep water warm. In this case, goggles must be worn by the students and beaker tongs must be available. For station 3, surface area, I originally had students add the tablets/powder to the water instead of the other way around, but the powder tended to sit on top of the water and ruin the result. Pouring the water onto the powder helps achieve the desired effect, but the beaker must be dry. I also recommend running each station on a tray to save cleanup of the lab benches or tables. I generally announce 12 minutes per station and let it run until 15 minutes if needed. Station 4 goes quickly, so I guide those students to look at the analysis section. Students may need extra time to complete the analysis questions.

To the class

In today's lab, we're testing out the effects of different variables on rates of chemical reactions. You will move between four stations, and you will have about twelve minutes per station. At station 1, you'll test temperature. You can obtain your warm water sample at the faucet [or the hot plate if needed]. There is also ice available. You will only need a few ice cubes for your trial. There is a lot of vinegar used in this lab, so the scent will permeate the room.

Discussion

Temperature has a direct relationship on rate of reaction. Therefore, higher temperatures lead to faster rates. This is because particles are traveling faster, which leads to more collisions.

Concentration also has a direct relationship on rate of reaction. Therefore, higher concentration leads to faster reaction rates. This is because there are more materials available for collisions.

Surface area has a direct relationship on rate of reaction. Therefore, more surface area leads to faster reaction rates. This is because there are more reaction sites and more places for molecules to collide.

Nature of reactants has a variable effect on the rate. It all comes down to what's reacting.

Effect of a catalyst increases the rate of reaction. A catalyst lowers the activation energy of a reaction, therefore allowing the reaction to proceed with less energy. This allows the reaction to occur faster. This was not tested in this lab.

Names Period Date Teacher

Reaction Kinetics Lab

Purpose: Chemical reactions take place when two substances come together under specific conditions. When some of these conditions change, the rate of the reaction can also change. In this lab, you will explore four different factors that affect the rate of reaction.

Station 1: Temperature

Materials: vinegar (300 mL), thermometer, baking soda, beakers, weighing tray, balance, stopwatch

Procedure

1) Obtain some ice. 5-10 cubes will do. Place this in an empty beaker and add vinegar until it reaches 50 mL. Stir until the temperature no longer changes.
2) For the iced vinegar, do the following. (Read all the steps before carrying them out.)
   a) Weight out 5 grams of sodium bicarbonate (or 1 heavy scoop)
   b) Find the temperature of the liquid and record it.
   c) Prepare your stopwatch.
   d) Drop the baking soda into the sample and simultaneously start the stopwatch.
   e) Watch until the reaction is finished, then stop the stopwatch.
   f) Record the time as well as the bubble height.
3) Repeat step 2 for room temperature vinegar.
4) If available, obtain a 50 mL sample of warm vinegar and run step 2.
5) If time permits, run a second trial for each.
6) Once all samples have been tested, empty and rinse the beakers in the sink. You do not need to dry them.

Vinegar Type	Trial 1			Trial 2 (Do all Trial 1 first)		
	Temperature	Reaction Time	Bubble Height (mL)	Temperature	Reaction Time	Bubble Height (mL)
Iced Vinegar						
Room Temperature Vinegar						
Warm Vinegar						

Station 2: Concentration

Materials: graduated cylinder, 250 mL beaker, vinegar (300 mL), baking soda, stopwatch

Procedure Part 1:

1) Measure out 25 mL of vinegar in the graduated cylinder and pour it into the beaker.
2) Measure out 50 mL of water in the graduated cylinder and add it to the beaker.
3) Weigh out 5 grams of baking soda (or 1 heavy scoop).
4) Prepare your stopwatch.
5) Drop the baking soda into the mixture and start the stopwatch.
6) Watch until the reaction finishes and record the time in the 33% Vinegar Concentration box. Also record how high up the bubbles went (approximations are okay).
7) Rinse out the mixture beaker and the graduated cylinder.

Procedure Part 2:

1) Measure out 50 mL of vinegar in the graduated cylinder and pour it into the beaker.
2) Measure out 25 mL of water in the graduated cylinder and add it to the beaker.
3) Weight out 5 grams of baking soda (or 1 heavy scoop).
4) Prepare your stopwatch.
5) Drop the baking soda into the mixture and start the stopwatch.
6) Watch until the reaction finishes and record the time in the 66% Vinegar Concentration box. Also record how high up the bubbles went (approximations are okay).
7) Rinse out the mixture beaker and the graduated cylinder.

Procedure Part 3:

1) Measure out 75 mL of vinegar in the graduated cylinder and pour it into the beaker.
2) Do not add water.
3) Measure out 5 grams of baking soda (or 1 heavy scoop).
4) Prepare your stopwatch.
5) Drop the baking soda into the beaker and start thc stopwatch.
6) Watch until the reaction finishes and record the time in the 100% Vinegar Concentration box. Also record how high up the bubbles went (approximations are okay).
7) Rinse out the mixture beaker and the graduated cylinder.

Mixture	Vinegar Concentration	Trial 1		Trial 2 (Do all Trial 1 first)	
		Reaction Time	Bubble Height (mL)	Reaction Time	Bubble Height (mL)
25 mL vinegar 50 mL water	33%				
50 mL vinegar 25 mL water	66%				
75 mL vinegar	100%				

Station 3: Surface Area

Materials: 250 mL beaker, 3 antacid tablets, weighing tray, stopwatch, mortar and pestle, graduated cylinder

NOTE: When dealing with antacid tablets, the reaction is considered complete when the tablet has dissolved. There may still be some bubbling, but you can ignore that.

Procedure Part 1:

1) Measure 100 mL water into the graduated cylinder.
2) Make sure the reaction beaker is completely dry inside.
3) Place one antacid tablet in the beaker.
4) Prepare your stopwatch.
5) Pour the water into the beaker and record the reaction time.
6) Rinse and dry out the beaker.

Procedure Part 2:

1) Measure 100 mL water into the graduated cylinder.
2) Make sure the reaction beaker is completely dry inside.
3) Break one antacid tablet into four pieces. Place the pieces in the beaker.
4) Prepare your stopwatch.
5) Pour the water into the beaker and record the reaction time.
6) Rinse and dry out the beaker.

Procedure Part 3:

1) Measure 100 mL water into the graduated cylinder.
2) Make sure the reaction beaker is completely dry inside.
3) Crush the antacid tablet into a powder or small pieces. Place the powder in the beaker.
4) Prepare your stopwatch.
5) Pour the water into the beaker and record the reaction time.
6) Rinse and dry out the beaker.

Tablet	Surface Area	Trial 1		Trial 2 (Do all Trial 1 first)	
		Reaction Time	Bubble Height (mL)	Reaction Time	Bubble Height (mL)
Full Tablet	Small				
Quarter Tablets	Medium				
Powder	Large				

Station 4: Nature of Reactants

Materials: baking soda, lemon juice (100 mL), vinegar (100 mL), water, 250 mL beaker, graduated cylinder, stopwatch

Procedure:

1) Choose one liquid sample to test.
2) Measure out 50 mL of the sample and pour it into the beaker.
3) Weight out 5 grams of baking soda (or 2 scoops).
4) Prepare your stopwatch.
5) Drop the baking soda into the beaker and record the reaction time. Also record the bubble height.
6) Rinse out the reaction beaker and the graduated cylinder.
7) Repeat for the other liquid samples.

Liquid	Liquid Type	Trial 1		Trial 2 (Do all Trial 1 first)	
		Reaction Time	Bubble Height (mL)	Reaction Time	Bubble Height (mL)
Vinegar	Stronger Acid				
Lemon Juice	Weaker Acid				
Plain Water	Neutral				

Analysis

1. Sketch a graph for each of the four stations. You may use line graphs or bar graphs.

Station 1

Temperature
vs.
Reaction Time

Station 2

Concentration
vs.
Reaction
Time

Station 3

Surface Area
vs.
Reaction Time

Station 4

Acidity
vs.
Reaction
Time

NOTE: Rate of reaction is the INVERSE of time. A shorter time has a faster rate of reaction.

2. What effect did temperature have on the rate of reaction?

3. What effect did concentration have on the rate of reaction?

4. What effect did surface area have on the rate of reaction?

5. What effect did the nature of the reactants have on the rate of reaction?

6. Which station best showed an effect on the rate of reaction? (i.e.: Which had the greatest variation in reaction times?)

7. Science says temperature should have an effect on the rate of reaction in that as temperature increases, the rate of reaction increases. Did this hold true for the experiment? If not, why might it have not worked?

8. Science says concentration should have an effect on the rate of reaction in that as concentration increases, the rate of reaction increases. Did this hold true for the experiment? If not, why might it have not worked?

9. Science says surface area should have an effect on the rate of reaction in that as surface area increases, the rate of reaction increases. Did this hold true for the experiment? If not, why might it have not worked?

10. Science says the nature of reactants should have an effect on the rate of reaction. Acids and bases react to neutralize each other. Water is considered neutral and is neither an acid nor a base. If this is the case, what type of compound is an antacid tablet, based on your data.

11. Antacid tablets are mostly comprised of sodium bicarbonate (baking soda), but it has other chemicals that cause it to react with water, including citric acid. When a tablet contacts water, the citric acid and sodium bicarbonate react, creating the fizzing effect. How might this reaction have affected the experiment?

12. Choose one station and describe how to make changes to it so it works better at showing reaction kinetics.

13. What are some experimental errors that could have taken place during the experiment? Provide three examples.

Conclusion

Hopefully this collection of activities has been useful. If so, please leave a review so other classroom teachers and homeschool teachers can find it. Reviews help authors increase exposure, so each one counts. You can access the review page by navigating through red-jade.com/steam.

If you have questions, suggestions, or want to report errata, you can reach me via steve@stephenjwolf.com

WORKS BY STEPHEN J. WOLF

Red Jade Fantasy Series
Red-Jade.com

Two kingdoms are in constant conflict, but one mage is determined to stop it. By finding and uniting the shards of the mystical Red Jade, Dariak hopes to find the power he needs to end the war. On his journey, he meets new companions, battles fierce monsters, and faces challenges at every turn. Will uniting the Red Jade bring peace or cause problems of its own?

Book 1:
Journeys in Kallisor

Book 2:
The Shattered Shards

Book 3:
The Assembly

Book 4:
The Forgotten Tribe

Book 5:
Delminor's Trials
A Stand-Alone Prequel

Book 6:
Dariak's Shadow
A Stand-Alone Prequel

Red Jade Omnibus
Books 1-4

red-jade.com/omnibus

The main four books of the Red Jade series gathered into one massive volume. Available in hardcover, on Kindle, and in the Apple Store.

Red Jade
Art Collection

red-jade.com/artcollection

All the artwork for the Red Jade series gathered in one place, including character sketches, covers, maps, and a bestiary, all in full color.

Kershin the Fire Mage

red-jade.com/kershin

Without control over his power, Kershin brings destruction with him.
Hessia is determined to stop him.

Legend of the Starsword

red-jade.com/starsword

Startok and Anthroff set out on a journey to save the dying land with magic and a special sword, meeting companions along the way and facing danger at every turn.

Lord of Time
(Starsword Book 2)

red-jade.com/lordoftime

A curse befalls the land and it must be reversed, but only Andrew suspects how. With Starlonn, Taniata, and Kimdeylyn, they must do the unthinkable to save the realm.

Last of the Lortens
(Starsword Book 3)

red-jade.com/lortens

As magic in the land fades, Andeson struggles with his future role as Lord of Timlefont. Cavish, a Knight of the Order, crosses his path and changes his destiny.

Coding for Kids:
Learn JavaScript:
Build
The Room
Adventure Game

red-jade.com/cfk1

Learn JavaScript while creating an adaptable text adventure game that challenges the player to locate items in a house and to escape by finding pieces of a passcode.

Coding for Kids:
Learn JavaScript:
Build
Mini Apps

red-jade.com/cfk2

Learn JavaScript while creating fifteen unique apps. From a word replacement game to tic-tac-toe to Battles in Kallisor, these apps teach you proper coding skills.

Coding for Kids:
Learn JavaScript:
Build
The Catch the
Culprit Game

red-jade.com/cfk3

Learn JavaScript while creating a detective-style game, where you find clues to catch the person who committed a crime. This is an intermediate level book.

Enzor
and the Volcano
The Book of Fire
A Choose-the-Fate Story

red-jade.com/enzor

Enzor quests to become a master magician, but he must face the perils of a volcano, and he needs your help to succeed.

Calla
and the Forest
The Book of Earth
A Choose-the-Fate Story

red-jade.com/calla

Calla must travel a forest to gather herbs to help her sick mother. She needs your help choosing which herbs to take.

Laric
and the Castle
The Book of Air
A Choose-the-Fate Story

red-jade.com/laric

Laric yearns to find the magisters that were taken from his village. He needs your help to navigate the castle.

Jenia and the Storm

The Book of Water

A Choose-the-Fate Story

red-jade.com/jenia

A typhoon threatens Jenia's village. Only use of magic crystals can protect it, but they're scattered among the islands. Jenia needs your help to find them.

Garinor's Adventure

A Choose-the-Fate Novel

red-jade.com/garinor

Garinor is taken from his home, but he doesn't know why. Soon after, things go awry, and he needs your help to guide him to his destiny.

Of Ages Past: A Poetry Collection

red-jade.com/ofagespast

94 poems about love, friendship, loss, science, dedications, and betrayal.

A Shocking Journey

red-jade.com/shocking

Learn the fundamentals of electricity and magnetism using analogies in this short story.

Magic and Code United

red-jade.com/magicandcode

The true story of our wedding season, from setting up table arrangements to the big day. Full color.

A keepsake for those involved.

STEAM Laboratory Activities

Focus on Chemistry

red-jade.com/steam

A collection of chemistry-related activities for use in school or at home.

Made in the USA
Monee, IL
07 November 2026

67974634R00111